EVENTS THAT
CHANGED THE
WORLD

1880–1900

═══The Nineteenth Century═══

**Other books in the
Events That Changed the World series:**

EVENTS THAT CHANGED THE WORLD

1880–1900

The Nineteenth Century

Jodie L. Zdrok, *Book Editor*

Bruce Glassman, *Vice President*
Bonnie Szumski, *Publisher*
Helen Cothran, *Managing Editor*

GREENHAVEN
PRESS ®

THOMSON
GALE

San Diego • Detroit • New York • San Francisco • Cleveland
New Haven, Conn. • Waterville, Maine • London • Munich

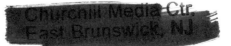

© 2004 by Greenhaven Press. Greenhaven Press is an imprint of Thomson Gale, a part of the Thomson Corporation.

Thomson is a trademark and Gale [and Greenhaven Press] are registered trademarks used herein under license.

For more information, contact
Greenhaven Press
27500 Drake Rd.
Farmington Hills, MI 48331-3535
Or you can visit our Internet site at http://www.gale.com

Cover credit: © Bettmann/CORBIS
Library of Congress, 27, 74, 110, 157

LIBRARY OF CONGRESS CATALOGING-IN-PUBLICATION DATA

1880–1900 / Jodie L. Zdrok, book editor.
   p. cm. — (Events that changed the world)
Includes bibliographical references and index.
ISBN 0-7377-2037-9 (lib. : alk. paper)
   1. World politics—To 1900. 2. Triple Alliance, 1882. 3. Technological innovations. 4. Inventions. 5. Hitler, Adolf, 1889–1945. 6. China—History—Boxer Rebellion, 1899–1901. I. Zdrok, Jodie L. II. Series.
D397.A15 2004
909.81—dc22   2003067762

Printed in the United States of America

CONTENTS

tions demanded that the Triple Alliance be formed according to the diplomatic and strategic considerations of the time.

Event 3: The International Meridian Conference Sets Greenwich Mean Time: October 13–22, 1884

Event 4: The Berlin Conference Divides Africa: February 26, 1885

Event 5: Louis Pasteur's Rabies Vaccine Is Successful: July 6, 1885

Event 11: Marie and Pierre Curie Discover Radium: December 26, 1898

1. The Curies Did Revolutionary Research

2. For the Beauty of Science

Event 12: The Boxer Rebellion in China Stirs the World: June 20–August 14, 1900

1. Antiforeign Sentiment in China

2. A Missionary in the Midst of the Uprising

FOREWORD

In 1543 a Polish astronomer named Nicolaus Copernicus published a book entitled *De revolutionibus orbium coelestium* in which he theorized that Earth revolved around the Sun. In 1688, during the Glorious Revolution, Dutch prince William of Orange invaded England and overthrew King James II. In 1922 Irish author James Joyce's novel *Ulysses*, which describes one day in Dublin, was published.

Although these events are seemingly unrelated, occurring in different nations and in different centuries, they all share the distinction of having changed the world. Although Copernicus's book had a relatively minor impact at the time of its publication, it eventually had a momentous influence. The Copernican system provided a foundation on which future scientists could develop an accurate understanding of the solar system. Perhaps more importantly, it required humanity to contemplate the possibility that Earth, far from occupying a special place at the center of creation, was merely one planet in a vast universe. In doing so, it forced a reevaluation of the Christian cosmology that had served as the foundation of Western culture. As professor Thomas S. Kuhn writes, "The drama of Christian life and the morality that had been made dependent upon it would not readily adapt to a universe in which the earth was just one of a number of planets."

Like the Copernican revolution, the Glorious Revolution of 1688–1689 had a profound influence on the future of Western societies. By deposing James II, William and his wife, Mary, ended the Stuart dynasty, a series of monarchs who had favored the Catholic Church and had limited the power of Parliament for decades. Under William and Mary, Parliament passed the Bill of Rights, which established the legislative supremacy of Parliament and barred Roman Catholics from the throne. These actions initiated the gradual process by which the power of the government of England shifted from the monarchy to Parliament, establishing a democratic system that would be copied, with some variations, by the United States and other democratic societies worldwide.

Whereas the Glorious Revolution had a major impact in the political sphere, the publication of Joyce's novel *Ulysses* represented a revolution in literature. In an effort to capture the sense of chaos and discontinuity that permeated the culture in the wake of World War I, Joyce did away with the use of straightforward narrative that had dominated fiction up to that time. The novel, whose structure mirrors that of Homer's *Odyssey*, combines realistic descriptions of events with passages that convey the characters' inner experience by means of a technique known as stream of consciousness, in which the characters' thoughts and feelings are presented without regard to logic or narrative order. Due to its departure from the traditional modes of fiction, *Ulysses* is often described as one of the seminal works of modernist literature. As stated by Pennsylvania State University professor Michael H. Begnal, "*Ulysses* is the novel that changed the direction of 20th-century fiction written in English."

Copernicus's theory of a sun-centered solar system, the Glorious Revolution, and James Joyce's *Ulysses* are just three examples of time-bound events that have had far-reaching effects—for better or worse—on the progress of human societies worldwide. History is made up of an inexhaustible list of such events. In the twentieth century alone, for example, one can isolate any number of world-shattering moments: the first performance of Igor Stravinsky's ballet *The Rites of Spring* in 1913; Japan's attack on Pearl Harbor on December 7, 1941; the launch of the satellite *Sputnik* on October 4, 1957. These events variously influenced the culture, society, and political configuration of the twentieth century.

Greenhaven Press's Events That Changed the World series is designed to help readers learn about world history by examining seemingly random events that have had the greatest influence on the development of cultures, societies, and governments throughout the ages. The series is divided into sets of several anthologies, with each set covering a period of one hundred years. Each volume begins with an introduction that provides essential context on the time period being covered. Then, the major events of the era are covered by means of primary and secondary sources. Primary sources include firsthand accounts, speeches, correspondence, and other materials that bring history alive. Secondary sources analyze the profound effects the events had on the world. Each reading is preceded by an introduction that puts it in context and emphasizes the event's importance in the ongoing evolution of world history. Additional features add to the value of the series: An annotated table of contents and an index allow readers to quickly locate material of interest. A chronology provides an easy reference for contextual information. And a bibliography offers opportunities for further exploration. All of these features help to make the Events That Changed the World series a valuable resource for readers interested in the major events that have shaped the course of humanity.

Discovery and Innovation in the Late Nineteenth Century

"**L**et there be light," cued inventor Thomas A. Edison one October evening in 1883. Upon Edison's spoken words, the village center of Brockton, Massachusetts, lit up with newly installed electrical streetlights. The test of the world's first underground, standardized central power system proved successful. This method of lighting became the cornerstone power-supply method throughout the nation.

The moment was not just a triumph of technological achievement, it had symbolic meaning as well. The effect of the scientific discoveries and technological innovations in the closing decades of the nineteenth century can itself be considered "illuminating." The late 1800s were years of great change.

The end of the century was characterized by numerous accomplishments in science and technology. These discoveries and inventions had a lasting impact during this period and are still among the late nineteenth century's most enduring legacies to the modern world.

Science and technology are overlapping spheres that complement and influence one another. Understanding facets of the natural and physical worlds is conducive to the keen, creative thinking that results in invention. For example, German physicist

Heinrich Hertz proved the existence of electromagnetic waves in 1888, and this led directly to Italian physicist Guglielmo Marconi's invention of the radio in 1895. This building-block nature of discovery and invention means that advances in one area led directly to developments in another. Several cultural and scientific trends prominent at the end of the century illustrate their symbiosis.

During this period education expanded to include many sectors of the populace. In England, the Education Act of 1870 provided for public education throughout England and Wales. Higher education also expanded during this period in England: Six universities were founded between 1880 and 1900. These broadened educational opportunities typify the changes taking place throughout the nineteenth-century Western world. Such improved opportunities for education allowed for basic scientific knowledge to become widespread. This, in turn, widened the pool of potential scientists and inventors. In addition, a more educated general public was better able to understand and appreciate the advances made in science and technology.

By the end of the century, the field of science had been transformed through increasing professionalization. Specialists emerged as scientists worked on specific areas, such as microbiology (Louis Pasteur) and physics (Marie and Pierre Curie). Specialization in medical science was especially prevalent. Greater concentration in specific fields of medicine enabled practitioners to devote more time and research to specific issues or questions. Historian George Weisz concludes, "By the end of the nineteenth century, at the latest, medical science had, largely as a consequence of specialization, developed to the point where the impossibility of mastering all of it seemed obvious."[1] This idea may be applied to specialization in all the sciences. A wish to increase knowledge resulted in precisely that, with the corollary that scientific knowledge became so expansive and specialized that "science" came to embrace many disciplines. Faster progress was therefore made in certain areas, such as physics and medicine.

Science and Climate Yield Invention

Technology, too, had a great impact on everyday life by the end of the nineteenth century. As the partner of science, technology often put new discoveries into practice through inventions. The plethora of new inventions in the last two decades of the nineteenth century had an effect on the way people communicated with one another, received information, and passed their leisure time.

What These Advances Meant for Society

Scientific advances and technological innovations had an impact on the way people lived. Anthrax and rabies vaccines, radio waves and X-rays are but a sample of the scientific discoveries made in the last quarter of the 1800s. During that same span of time, the telephone, light bulb, motion-picture camera, and radio were invented, to name but a few innovations at the end of the century.

Science and technology also modified ways of thinking. Late-nineteenth-century society granted science authority. Many people believed that science and technology held the answers to the problems of the world. Discoveries and inventions promoted a hopeful, positive view of progress. The late nineteenth century instilled in many people the promise of a new, exciting age of science and technology. The general view was that the world was becoming better and better. Industries took pride in their pursuits, whether it was shipbuilding companies seeking to build the biggest and best ships ever or factory owners pushing for the most efficient way to produce materials.

And science *did* improve people's lives. Although disease was not fully eradicated, as many supposed it would be, advances improved the quality and length of life. Communication and transportation became accessible, and the world became "smaller" as global connections became more immediate. Scientists facilitated professional and popular understanding of physical and natural laws, and inventors used the new insights of science to create helpful or time-saving products. Science and technology accomplished what seems to be characteristic of the era: a reflection of the past and a preparation for the future.

Note

1. George Weisz, "The Emergence of Medical Specialization in the Nineteenth Century," *Bulletin of the History of Medicine*, 2003, p. 538.

**President James Garfield Is Assassinated:
July 2, 1881**

Garfield's Assassination Spurs Civil Service Reform

by Sean Dennis Cashman

In nineteenth-century America, politicians rewarded campaign sup-
porters with jobs and political positions through a practice called
the spoils system. Party loyalty, rather than merit and skill, was the
basis for civil service appointments. Political parties had debated
the need for civil service reform since the 1860s, but it was neither a
priority nor an easy problem to address. The spoils system had been
in effect for many years, and many politicians at least partly owed
their careers to the system. A tragedy then took place: On July 2,
1881, just a few months after his inauguration, President James A.
Garfield was shot by a man who was distressed about having been
overlooked for a political patronage job. Garfield died several
months later from the gunshot wound.

One effect of Garfield's assassination was to make civil service
reform an imperative. Chester A. Arthur, successor to Garfield,
pressed forward with reform despite his own political appointment
through the spoils system. On January 16, 1883, Congress passed
the Pendleton Act, which regulated federal civil service positions
and placed them under the jurisdiction of a special commission.
Candidates for government jobs were now required to pass exams,

Sean Dennis Cashman, *America in the Gilded Age: From the Death of Lincoln to the Rise of
Theodore Roosevelt*. New York: New York University Press, 1984. Copyright © 1984 by New
York University Press. Reproduced by permission.

and changes in presidential administrations no longer affected a person's job security. The Pendleton Act set the precedent for a merit- and skill-based civil service system in the United States. However, a spoils system still exists—to a certain extent.

In this article Sean Dennis Cashman explains the spoils system and its relationship to civil service reform. He also traces the different political perspectives on the reform issue as well as the political atmosphere surrounding the Garfield presidency.

Cashman is the author of eight books on U.S. history. In addition to writing about late-nineteenth-century America, he has also written about the country's history during the 1920s and 1930s.

P residents held official power for short periods. On the other hand, the party bosses who nominated them wielded absolute power for as long as a generation. State bosses might be governors or senators. If they were not, they nominated those who were. Their power base was either considerable financial or industrial interests or control of a crucial area of the civil service.

The Spoils System and Civil Service Reform

The American civil service was part of a spoils system in which federal patronage was dispensed by party bosses. Officeholders were selected on the basis of party loyalty, not administrative competence. It was standard practice for new administrations to nominate their own appointees. The distribution of spoils after an election victory was called by Matthew Quay of Philadelphia, "shaking the plum tree." The climax came when [President Abraham] Lincoln, who had 1,639 places in his gift, removed 1,457 officeholders. By the early 1870s the civil service was made up of incompetent and demoralized party hacks. As A.C. McLaughlin explains in *The Courts, the Constitution, and Parties* (1912), the spoils system "provided a means of financing party management, it furnished the sinews of war to party government.". . .

Reform of the civil service was one of the few subjects on which both liberals and conservatives could agree. In the wake of new inventions, industry, and technology, it was becoming clear to most people that nothing short of professional expertise would do if the government were ever to govern a nation undergoing momentous industrial and economic transformation. The old spoils system would have to go. Representative Thomas A. Jenckes of Rhode Island had begun to agitate for selection by competitive examination as early as 1865. . . .

Some politicians were genuinely opposed to a civil service selected on the basis of academic attainment rather than practical ability. To James B. Beck of Kentucky and others the existing system represented a natural school of citizenship. As E.L. Godkin, editor of the *Nation*, observed, it was possible that a system of competitive examinations would discriminate against able men from poor backgrounds who could not afford to go to college. Thus, instead of weakening class prejudice it would have the opposite effect. A nonentity like Franklin Pierce could become president while a genius like Lincoln could not. . . .

Garfield as President and as Victim

Garfield was the last of the seven log cabin presidents beginning with Andrew Jackson. A self-taught scholar who could speak both German and French and write Latin with one hand and Greek with the other at the same time, he nevertheless found presidential duties beyond him. "My god! What is there in this place that a man should ever want to get in it," he exclaimed after a month in office. Garfield and [U.S. representative James G.] Blaine had been friends since Blaine had protected Garfield from disgrace in the Crédit Mobilier scandals [scandals involving railroad stock] of 1872. Now Garfield needed Blaine more than ever. He had been decisive in making the president; he was stronger in the party. It was inevitable that he would take the leading post in the cabinet, that of secretary of state. Garfield was disposed to conciliate faction by compromise. Blaine wanted his rivals killed off.

The collectorship of the New York Custom House was the most important administrative post outside the cabinet. It now controlled 1,300 apointments. It was there that Garfield and Blaine struck first. Rather than propitiate the Stalwarts [a faction of the Republican Party that opposed civil service reform], Garfield affronted them by appointing W.H. Robertson as collector. Robertson was not only one of Blaine's allies but also another adversary of [U.S. senator Roscoe] Conkling. Conkling and his ally, Senator Tom Platt of New York, retaliated. At first they blocked Senate approval of Robertson and then on May 16, 1881, resigned their seats in protest at Garfield's "perfidy without parallel." Conkling and Platt confidently expected that the New York legislature at Albany would reappoint them. But the Half Breeds [a moderate faction of the Republican Party that supported moderate civil service reform] discovered Platt in an Albany hotel in the company of "an unspeakable female." He was so embarrassed that he withdrew his candidacy. Conkling was not reelected. The very day on which the public was regaled with newspaper reports of Platt's sex life farce turned to tragedy.

On July 2, 1881, James Garfield had breakfast with the Blaines and left their house for the Baltimore and Potomac Railroad Depot in Washington. He was bound for a college reunion at Williamstown, Massachusetts. But at the depot he was shot and fatally wounded by a disappointed office seeker, Charles Guiteau. As he fired his pistol into the president's back he cried, "I am a Stalwart and Arthur is president now." [The assassin's identification with Stalwarts ended the use of these labels.] Contrary to expectations, Garfield did not die immediately. His life hung in the balance, and he lingered for three months. The wound hardly bled and the doctors had great difficulty in locating the bullet, which had lodged in muscle. Among those who tried unsuccessfully to do so was inventor Alexander Graham Bell, who used an electrical device. Ironically, it was not the bullet that killed the president but the efforts to save him. The inexperienced surgeons were clumsy, and their careless probes caused blood poisoning, from which Garfield died on September 19, 1881.

Agitation for Civil Service

"Martyrdom is the only way a man can become famous without ability" was a cynical observation of George Bernard Shaw's. Indeed, the assassination of Garfield in 1881 brought the agitation for civil service reform to a head in much the same way as the assassination of John F. Kennedy in 1963 gave momentum to the movement for civil rights. While Garfield's life ebbed slowly away people became increasingly aware of the way quarrels over patonage had led to his assassination. They were not even distracted by morbid press accounts of the trial and the hanging of his assassin. On August 11, 1881, the National Civil Service Reform League was founded.

The new president surprised everyone. Chester A. Arthur astonished his admirers and angered his adversaries by abandoning the spoils system that had made him president. He vetoed lavish appropriations of $18.7 millions for a Rivers and Harbors bill in 1882 because it would have provided new opportunities for graft and corruption. And on May 15, 1882, he gave his assent to an act establishing a commission to report on tariff revision.

Arthur's life in the demimonde had made him a good judge of character. His nights spent in smoke-filled rooms had been a political education in themselves. . . . Arthur wore the pants in his own home. In fact, he was quite a dude. He had eighty pairs of them. His taste was impeccable. The dude president invited Associated Artists, the company of interior decorators founded by artist and glass designer Louis Comfort Tiffany, to modernize the White House.

Scandals and Reform

Arthur's political resolve was strengthened by public indigation over two new scandals, the Star Route Frauds and Hubbel Hale's campaign letter. In early 1881 Postmaster General Thomas L. James had the postal service investigated by James MacVeagh, who exposed malpractices in the western mail services. In the sparsely populated West, contracts for mail delivery were awarded to private firms that used stagecoaches and horseback riders to carry letters and packets, since train and steamboat were not available. These firms were supposed to deliver mail with "Certainty, Celerity, and Security," the words emphasized on the contracts by three asterisks or stars.

It was the duty of Thomas J. Brady, second assistant postmaster general from 1876, to administer the routes and award the contracts. MacVeagh discovered that Brady had not only given contracts in exchange for bribes but had also accepted outrageous charges from dishonest contractors. The total cost to the government was $4 million. Brady and his confederates were arraigned and tried. A first trial ended inconclusively on September 11, 1882, and a second in a surprising acquittal on June 14, 1883.

The second scandal cut to the heart of the matter of civil service reform, the need for a service independent of party. In May 1882 the Chairman of the Republican congressional campaign committee, Jay A. Hubbel Hale, sent to all civil servants the traditional circular letter asking them for "voluntary contributions" to the Republicans' election campaign funds. According to the usual custom, campaign contributions were between 2 and 6 percent of salaries. In the atmosphere following Garfield's death and the unsuccessful prosecution of the Star Route Frauds, the letter inflamed public opinion. Public hostility to the Republicans was registered in the election results of 1882. The Democrats carried both houses. They then promoted legislation to reform the civil service, sponsored by Senator George H. Pendleton of Ohio. Republicans had to prove to the electorate their willingness to support a cause for which a president had died if they wanted to remain in office.

The Pendleton Act, signed by Arthur on January 16, 1883, created a Civil Service Commission of three members appointed by the president with the consent of the Senate. Not more than two were to come from any one party. The commission was to appoint a corps of examiners and provide for competitive examinations for prospective federal employees. The examination regulations were to apply to government departments in Washington and to customhouses and post offices with more than 50 employees. The president could extend the provisions of the act to other offices. The solicitation of campaign contributions was forbidden. By the turn of the century almost 100,000 positions were on the classified list. It was not only fortunate but also essential that the merit principle was adopted and put into effect before the increased role of government in the twentieth century necessitated an expanded civil service. Without it the fundamental reforms of the progressives could not have been attempted.

Agitation for civil service reform had signified much more than public discontent over political spoils. It represented a fundamental challenge to government by clique and cabal. In the long battle the public had learned much about political strategy and tactics. The prevailing oligarchy was visibly shaken. But the new plutocracy was not. And when the public turned its attention to trusts and robber barons in the contemporaneous chain of trust and railroad legislation, it was not assured of victory. For reform changed the nature of corruption rather than the nature of politics. During the 1880s it became ever more insidious. The professional politician was to be upstaged by the political businessman.

A Nation Reacts in Sympathy

by William Ralston Balch

The assassination of President James Garfield on July 2, 1881, created a sense of urgency about civil service reform. The tragedy of this event also reveals the sentiments of a nation undergoing the shock of a presidential assassination. President Garfield did not die immediately from the bullet wound. He lingered in illness throughout the summer of 1881. Unsanitary surgical instruments and the inability of his doctors to locate the bullet proved to be fatal for the president. Blood poisoning and pneumonia set in, and the president's health declined further until his death on September 19, 1881. In his 1881 book *The Life of James Abram Garfield, Late President of the United States*, excerpted here, William Ralston Balch offers a portrait of the nation in the midst of this crisis.

William Ralston Balch was an international journalist and the author of several books.

After the first moments of amazement, incredulity, horror and suspense, the people—the world—offered condolence. The emotion and spectacle were without parallel. In every household there was a hushed and tender silence, as if one long loved lay dying. The public festivities of the Nation's birthday were stayed, and the crowds that had gathered to form festivals were transformed into praying congregations, earnestly petitioning the Throne of Grace for mercy for the President. Abroad, American gayety was given over. In the British Parlia-

William Ralston Balch, *The Life of James Abram Garfield, Late President of the United States*. Philadelphia: Hubbard Bros., 1881.

ment, Whig, and Tory, and Radical listened to catch from the lips of the Prime Minister, the latest tidings from the sufferer. From the French republic, from the old empire of Japan and the new kingdom of Bulgaria, from Parnell, the Irish agitator, and from the Lord Mayor of Dublin, came messages of sympathy and sorrow. Sovereigns and princes, the people and the nobles, joined in earnest hope for the life of the Republican President. The press of all Christendom told the mournful story, and moralized as it told. At home the popular grief was absolutely unanimous. One tender, overpowering thought called a truce even to party contention. Old and young, men and women of all nationalities and of all preferences, their differences forgotten, waited all day for news, watched the flags and every sign that might be significant, and lay down to sleep, thanking God that as yet the worst had not come.

Expressions of Sympathy

It was a marvelous spectacle. It was—this feeling of millions for that one man—profoundly touching. It blessed him with great distinction among mankind. It blessed the country in that it stirred the people with a great overmastering emotion. I have not space to chronicle all the words of sympathy that went on the wings of the wires to Washington. They would fill several volumes as large as this. In London, the shooting of the President excited the profoundest sensation of consternation and grief among the American residents. The greatest anxiety was everywhere manifested to hear further news. The offices of newspapers and news agencies were visited by crowds to gain information. Crowds gathered at the American Exchange in the Strand. As the news spread among the theatres and other places of resort, the Americans left the buildings, and many ladies and gentlemen, in evening dresses, went direct to the American Exchange for the latest details. There were numerous callers at United States Minister Lowell's private residence, to inquire concerning the President. Earl Granville, Foreign Secretary, received a telegram from the British Legation at Washington, announcing the sad affair, and he at once cabled his condolences. The Queen, who was at Windsor Castle, immediately, on receipt of the news, personally telegraphed to Minister Lowell a message expressing deep regret and concern. The Town Councils of many inland towns passed resolutions of sympathy. In the British Parliament allusions were made to the affair amidst profound silence and regret. The Poet Laureate telegraphed his condolences. The members of the Royal Family sent to our Minister to know all he knew. . . . Prayers were ordered daily in Westminster Abbey. The Lord Mayor of London telegraphed his condolences. [English prime minister William] Gladstone sent the following letter to Mrs. Garfield:

LONDON, July 21st, 1881.

DEAR MADAM: You will, I am sure, excuse me, though a personal stranger, for addressing you by letter to convey to you the assurance of my own feelings and those of my countrymen, on the occasion of the late horrible attempt to murder the President of the United States, in a form more palpable, at least, than that of messages conveyed by telegraph. Those feelings have been feelings, in the first instance, of sympathy, and afterward of joy and thankfulness almost comparable, and, I venture to say, only second to the strong emotions of the great nation of which he is the appointed head. Individually, I have, let me beg you to believe, had my full share in the sentiments which have possessed the British nation. They have been prompted and quickened largely by what, I venture to think, is the ever-growing sense of harmony and mutual respect and affection between the countries, and of a relationship which, from year to year, becomes more and more a practical bond of union between us; but they have also drawn much of their strength from a cordial admiration of the simple heroism which has marked the personal conduct of the President, for we have not yet wholly lost the capacity of appreciating such an example of Christian faith and manly fortitude. This exemplary picture has been made complete by your own contribution to its noble and touching features, on which I only forbear to dwell because I am directly addressing you. I beg to have my respectful compliments and congratulations conveyed to the President, and to remain, dear madam, with great esteem,

Your most faithful servant,
W.E. GLADSTONE.

To this Secretary [of State James] Blaine replied by cable:

WASHINGTON, July 22d, 1881.

LOWELL, Minister, London: I have laid before Mrs. Garfield the note of Mr. Gladstone, just received by cable. I am requested by her to say that among the many thousand manifestations of interest and expressions of sympathy which have reached her, none had more deeply touched her than the kind words of Mr. Gladstone. His own solicitude and condolence are received with gratitude. But far beyond this she recognized that Mr. Gladstone rightfully speaks for the people of the British Isles, whose sympathy in this national and personal affliction has been as quick and as sincere as that of her own countrymen. Her chief pleasure in Mr. Gladstone's cordial letter is found in the comfort which it brings to her husband. The President is cheered and pleased on his painful and weary way to health by the many messages of sympathy which, in his returning strength, he safely receives and most gratefully appreciates.

BLAINE, Secretary.

More Kind Words and Wishes

On the Continent, the head of every country hastened to offer sympathy. The Emperors of Russia, Austria, Germany; the Kings of Sweden and Norway, Denmark, Belgium, Portugal, Italy, and Spain; the Sultan of Turkey; the Presidents of France and Switzerland, hundreds of statesmen and distinguished men, sent, through various channels to Washington, kind words and wishes from sympathetic hearts.

At home, the universal expression of sympathy found appropriate channels in the governors of states, mayors of cities, legislatures, boards of trade, clubs, associations, and conventions of every description, grand juries, churches, etc., etc. The South particularly manifested a most noble sympathy. Governor R.W. Cobb, of Alabama, telegraphed:

> Reports of the favorable indications gladden the hearts of Alabamans, who profoundly sympathize with the President and his family, and bitterly denounce the cowardly and brutal attempt on his life. The great peril through which he is passing draws all men to him, and he will resume his duties with a more generous and patriotic support from the people of the whole country.

Congressman E.W. Robertson telegraphed from Baton Rouge, Louisiana:

> The heartfelt, outspoken sorrow of our people at the late dastardly attempt upon the life of the President, prompts me to express their prayerful hopes for his speedy recovery.

[Former Confederate president] Jefferson Davis wrote the following letter:

BEAUVOIR, MISS., July 5th, 1881.

MR. FINDLEY S. COLLINS—Dear Sir: I have received yours of the 4th inst., and thank you for the kind expressions it contained. The evil influences to which you refer as causing the bitterness felt toward Southern men, it may fairly be expected, will give way to the sober sense of the people, if they shall, like yourself, detect the sordid motives for which the stimulants are administered. I well like the telegram you cite in regard to the attempted assassination of the President. I am thankful the assassin was not a Southern man, but will say I regret that an American crime, black enough in itself, has a deeper dye from the mercenary motive which seems to have prompted it. I sincerely trust the President may recover, and that the startling event will arouse the people to the consideration of a remedy for the demoralization which a wild hunt after office is creating. With best wishes for your welfare, I am, very truly yours,

JEFFERSON DAVIS.

The Commander-in-Chief of the Grand Army of the Republic promulgated the following order:

HEAD-QUARTERS GRAND ARMY OF THE REPUBLIC,
BOSTON, July 7th, 1881.

[General Order, No. 42.]

Awaiting the fateful issue which hangs like a pall over our land, grateful for the glimpse of sunshine through the dark cloud, the soldiers of the Grand Army of the Republic tender their old comrade in arms, the stricken President, their sympathy and love. What lies behind the veil of the future we may not seek to know; but remembering that the same Almighty which guided us to victory is beside our fallen comrade, let us cast out all our fears and send to the throne of grace, not a cry of despair, but a prayer of hope and faith in the Divine wisdom and love.

(Signed),
GEO S. MERRILL, Commander-in-Chief.
WILLIAM M. OLIN, Adjutant-General.

The bench took notice of the national calamity. Said Judge Ludlow, of Philadelphia, in charging the Grand Jury for the July Term:

> We meet to-day under adverse circumstances; we are under the shadow of a great cloud, and our hearts beat with alternate hope and fear; we do not yet know what the end will be, but this we all understand, that law and order must prevail; that the constituted authorities must be respected and sustained, and woe betide the man who dares to raise even so much as his little finger against the integrity of the Republic or against the life or lives of its lawfully elected officers. I care not what may be your politics, faith or religious feelings. I know this, that as Americans, you represent not only the citizens of this country, but, in a sense, Americans everywhere all over this land, and you will join with me in as severe a condemnation of this anti-American crime as can be uttered in human language, and in a devout prayer to the Almighty that the life of the legally elected and inaugurated President of the United States may be spared to the nation and to his family.

James A. Garfield

The church was equally as ready. Prayers were offered in every pulpit in the land. Special services were held, imploring the Almighty to spare the President's life.

The Archbishop of Baltimore, the Most Rev. James Gibbons, issued the following to his clergy.

> REV. AND DEAR SIR.—You, in common with all others have heard with amazement and horror of the late attempted assassination of His Excellency, the President of the United States. It is scarcely possible to imagine a deed more appalling to men or more iniquitous before God; for if it is such a crime to slay even a private citizen, what an enormity is it to attempt the death of one who, while representing the whole nation, is also, as to matters temporal, the highest vicegerent of God Himself in the land; and the act of the assassin is the more heinous, since he had neither a private grievance to avenge nor the semblance of a public wrong to redress. And our detestation of the wretch who has stricken down our head is yet more increased when we add to the official dignity of the

sufferer his accessibility and affability to all, and his committing, like all his predecessors, his personal safety entirely to the good-will and good sense of those over whom he presides. Well may we stand aghast when, in this crime and in another like crime perpe-trated a few months ago, we see the mischief of which a single in-dividual is capable when he has ceased to fear God, to value man or to dread the consequences of giving free scope to his own pas-sion. In the fact, then, of this most hideous deed, we are called upon to express at once our loathing of the crime and our deep sympathy with him whom this crime has placed in such great suf-fering and such imminent peril, for while the Catholic Church is happily above all our parties, and is far from the wish to take to herself the decision of the very transient, and, as a rule, not very momentous questions as to which of these parties are at issue, yet none more than the Catholic Church inculcates respect for every duly constituted authority, or more reprobates or threatens every-thing by which such authority is assailed.

You will, therefore, with all the power at your command, urge our people to pray during Mass, and at other times, for the recovery of His Excellency, and on Sunday next, should he then still survive, you will say in his behalf, before or after Mass, and together with all your people, the Litany of the Saints, as at once entreating God to spare his life, and also as making an act of expiation for a crime which appertains to us as a nation, and not only concerns, but tar-nishes us all.

Very faithfully, your friend in Christ,
JAMES,
Archbishop of Baltimore.

In Kentucky and Arkansas the 14th of July was made, by proclamation, a day of fasting and prayer for the President's re-covery. Governor Foster of Ohio telegraphed the governor of every State, asking him to join in a movement to hold a day next autumn as a day of National Thanksgiving for the President's re-covery. Every governor except Governor Roberts of Texas, who is a crank, answered gladly. The movement of sympathy was in-deed universal from more than one hundred millions of men. It was to have been expected, as a matter of course, that those closely identified with him by long years of personal and politi-cal association, would, out of the fullness of their affection, min-gle their tears with those of his kindred; but that so eager inquiries and tender messages of sympathy should come from all over the world, is the most welcome evidence that all the world's akin. From every nook and corner of our land were messages sent freighted with loving regard, and even the cables, which lay beneath the ocean, were kept busy night and day transmitting the sympathy of the rulers, and princes, and peoples of all civilized nations of the globe.

The Resources of the Press

Here we will take a hasty glance at one of the remarkable features in the attempt on the life of the President, before continuing our direct narrative. I refer to the development of the extraordinary resources of the press at Washington. It came suddenly and without warning in the middle of the dull season. The winter force of correspondents had dwindled down one-half, and those who were left were dawdling away the first hours of the heated term in a semi-demoralized condition. Many newspapers had discharged their specials for the summer and were running shorthanded. Within twenty minutes on Saturday, the sleepy-looking reserve was thoroughly awake, had thrown out its pickets, patrols and videttes, and had begun the task of gathering in the details of the great crime and preparing it for the press, Recruits came flying down, twelve hours later, from New York, Philadelphia, Baltimore and other contiguous cities, and these, under the veterans of the Row, did veterans' duty. The cry of the home officers was "send unlimited." The response was in accordance therewith. Every newspaper man was worth two, three or four of ordinary occasions, for he not only developed twice the energy but worked all day and all night. The result was astonishing, and it is told with mathematical accuracy by the figures of the telegraph offices. On Saturday, alone, 275,000 words were wired from the Western Union office. This was the highest record of newspaper work ever done at the National Capital. On last Inauguration Day, 190,000 words were handled, the highest number up to the event of July 2d. But the press and telegraph companies had full warning and had made every preparation previous to the fourth of March. It was in a busy season, and a full corps of experienced men were at work. Previous to the Inauguration record, the highest score was during the Credit Mobilier excitement in 1873, and this was only 113,000 words. [The Credit Mobilier scandal of 1872–73 involved fraud, the federal government,

and the reputations of numerous politicians and businessmen.]
The highest record at the Capital on sensational days has not
gone over 50,000 words. It will be seen, then, that Saturday, July
2d, was an era in Washington correspondence and telegraphy,
both of which reached a point of excellence and capacity, under
the most disadvantageous circumstances, hitherto unknown. The
record of words at the Western Union does not, however, tell the
whole story. Several of the most prominent newspapers, and notably those which take the most matter, have private wires, and
each of these, alone, sent from twenty to thirty thousand words.
The second of July, 1881, will long be remembered as a black
spot in American history, but it will also be remembered as a
bright one in the annals of American journalism.

The Triple Alliance Solidifies European Power

by Imanuel Geiss

In the latter half of the nineteenth century, the practice of creating alliances defined European foreign policy. These political alliances were agreements among nations to support each other in the event of military attack.

Germany and Austria-Hungary established the secret Dual Alliance in 1879. German chancellor Otto von Bismarck maintained that security would be more steadfast with additional members. Italy joined the partnership three years later after a dispute with France. On May 20, 1882, the Triple Alliance was formed, with the treaty guaranteeing mutual support should any of the nations be attacked by Russia or France. In 1907 a rival alliance was formed when Russia, France, and Great Britain formed the Triple Entente.

This selection by German historian Imanuel Geiss traces the complicated path to the Triple Alliance. He outlines the relationships among Germany, Austria-Hungary, Russia, France, and Great Britain. Geiss concludes that the creation of the Triple Alliance amounted to a concentration of power in Central Europe.

Imanuel Geiss, professor of history at the University of Bremen in Germany, has published a number of works about German history in the nineteenth and early twentieth centuries.

T he conclusion of the Dual Alliance reopened the problem of alliances in Europe. [German chancellor Otto von] Bismarck had hoped to group other Powers around the Dual Alliance, in order to isolate Russia, because he thought her to be more dangerous in her first phase of Pan-Slav agitation than an isolated French Republic. As the relationship between Britain and Austria had always been good, Bismarck hoped that 'Austria will carry Britain' into his camp. As Bismarck feared that a possible result of the Dual Alliance would be Russia's declaration of war against Germany, he tried to secure Britain's backing. Germany's cautious sounding out of Britain's position was answered at the end of September 1879 by [British prime minister Benjamin] Disraeli's equally cautious reply. He stated that Britain would keep France and Italy neutral in case of a war between Germany and Russia over the oriental question, if Germany were to support Britain there.

The Dual Alliance and Other Powers

This was not exactly what Bismarck had expected from Britain. When the Russian ambassador, a few hours after hearing about the British reply, came to Bismarck with accommodating proposals of renewing the League of the Three Emperors of 1873, Bismarck no longer felt Germany depended on British support. When Disraeli became more forthcoming and offered an alliance with Germany against Russia, two weeks later, Bismarck apparently did not even bother to reply. He hoped that Germany would be able to cope with both Russia and France without British support. Bismarck's hesitation was explained by his fear that the Liberals under [former British prime minister William] Gladstone would win in forthcoming General Elections. If Gladstone returned to power it would be improbable that Britain would join an alliance against the French Republic for the conservation of Turkey, since Gladstone was attacking Turkey in his election speeches. In March 1880 a Liberal majority government under Gladstone returned to the House of Commons, thus confirming Bismarck's fears. The possibility of Britain joining Germany or the Dual Alliance in one form or another was thus forestalled for the foreseeable future.

Happier perspectives, however, seemed to be opened by offers made by Russia in Berlin on 27 September 1879, to renew the League of the Three Emperors, which had just been shattered by the oriental crisis. After prolonged negotiations to overcome Austria's reluctance to come to terms with Russia, the old relationship of the three conservative Empires was superficially reconstructed on 18 June 1881 by another Treaty of the Three Emperors. Bismarck had recommended this to his Emperor with three arguments: (i) peace between the three great monarchies would be secured for years, because Tsar Alexander III who succeeded his father after the assassination of Alexander II, was known to be reliable, and a rupture between the three monarchies 'would considerably shake the strength of the monarchical principle in Europe vis-a-vis the revolutionary one; (ii) the 'danger of a Franco-Russian coalition would be completely eliminated and France's peaceful disposition towards us will be virtually ensured'; (iii) the 'attempts of the anti-German war party in Russia to influence the decisions of the young Tsar' would be foiled by the pledges given by Alexander III. For Bismarck the new treaty made by the three conservative monarchies restored an ideal constellation. The two 'German' or 'mid-European' Empires were, compared with the first League of the Three Emperors, attached to each other with special links, i.e. the Dual Alliance.

The Treaty of 1881 was not an alliance, but contained only the agreement to remain neutral in the case of an attack by a fourth power and to try and settle all problems concerning the Balkans and Turkey. An additional protocol specified further points: Austria-Hungary reserved the right to annex Bosnia-Herzegovina at a suitable time. Russia was prepared to concede to Austria-Hungary the right of occupation, not of annexation of the Sanjak (district) of Novibazar. The three powers had nothing against a future unification of Bulgaria with East Rumelia, but not with Macedonia a delicate compromise between Russia and Austria. The special protocol mentioned the points which would lead to future crises in the Balkans: Bosnia-Herzegovina, the Sanjak of Novibazar, Bulgaria and Macedonia. The treaty was shattered by the Bulgarian crisis of 1885 and lasted no more than four years.

Alliances Considered

After the restoration of the League of the Three Emperors closer links with Britain were improbable due to the presence of Russia. The idea of extending the Dual Alliance would only be possible if medium and smaller-sized Powers were to join it. Bismarck did not actively encourage new allies, but several candidates came forward, e.g. Roumania in 1880 and Turkey and Italy in 1881. After the loss of Bessarabia to Russia at the Congress of Berlin, Roumania feared that Russia wished to create direct territorial communications with Bulgaria by the annexation of the Dobrudja. Roumania wanted to protect herself against such a loss by joining the Dual Alliance. During the negotiations leading to the Treaty of the Three Emperors in 1881, Bismarck hesitated to commit himself, but promised the support of the Dual Alliance if Roumania were threatened by Russia. The dynastic links with the Catholic line of Hohenzollerns on the Roumanian throne and the chance to block direct territorial communications between Russia and Bulgaria with Roumania's help were good enough reasons to work together with Roumania. It was only in October 1883 that Roumania was allowed to join the Dual Alliance. The political effect of this move, however, proved to be marginal in the long run. In 1914 Roumania remained neutral, for the same reasons as Italy, and entered the war on the side of the Entente in August 1916.

Applications from Italy and Turkey were more difficult to handle. Turkey was weakened through her internal structure and the danger of being attacked from outside. When the Turkish government asked Berlin for help to fight against France in North Africa in December 1881, Bismarck refused promptly and energetically. He told the Turks he 'was firmly convinced that France harboured no designs on Tripoli'. Even when the Turkish Empire became highly attractive to German 'Weltpolitik' [Germany's imperialist policy under Kaiser Wilhelm II, devised with the aim of making Germany a world power] after 1895–6, Germany did not enter into a formal alliance with Turkey until 2 August 1914, when the German government was desperately trying to pick up allies wherever they could be found.

A New Link with Italy

Bismarck's mixed feelings about the link between Italy, Germany and Austria-Hungary stemmed from different reasons, one of them being Italy's long coastline which made her vulnerable to attack by sea. A more fundamental reason was Germany's ambivalent position towards Italy for historical and ideological reasons. Although Italy had achieved national unity only a few years before Germany and although she had been a direct ally to Prussia in 1866 and an indirect one in 1870, Bismarck did not think

Italy worthy of becoming an ally of the German Empire because of her liberal character and parliamentary structure. Bismarck did not believe in the future of the Italian monarchy and he feared that 'an Italian Republic would go together with the French Republic'. When the Italian Ambassador to Berlin formally asked to be accepted into the Dual Alliance on 31 January 1882, Bismarck pointed out that Italy's parliamentary constitution was an obstacle to be overcome. Giving Britain as an example, Bismarck explained that such a parliamentary structure would not guarantee a continuously conservative foreign policy. Bismarck also stressed that 'Italy offered . . . the spectacle of cabinets drifting continuously to the Left'. When the Ambassador assured him that his king could firmly rely on the army to pursue a policy acceptable to Bismarck, the German Chancellor pushed his conservative flippancy so far as to 'allow himself', as he put it, the ironical 'hint that a Monarch who is wearing civilian clothes does not do everything in his power to keep in contact with his army'. It was difficult for the German military monarchy where not only the Emperor but even his first ministers often appeared in military uniform in public, to place confidence in a liberal and parliamentary monarchy, with a king at its head who preferred civilian clothes to military uniforms.

Bismarck was not completely hostile. After humiliating the Italian ambassador Bismarck passed on the Italian request to Vienna. Because of the Italian irredenta in Austria (Trentino) and competing interests on the Adriatic Coast and in the Balkans, Italy and Austria-Hungary had to be organized before the Dual Alliance could be enlarged. Bismarck saw the strategical advantage of having Italy allied against France, while Italy was anxious to join Germany and Austria-Hungary for reasons of foreign and domestic policy.

Triple Alliance Provisions

The Triple Alliance Treaty was signed on 20 May 1882, without great enthusiasm on the part of Austria-Hungary and Germany. The Triple Alliance would, however, give Germany more power and prestige than the Dual Alliance, which remained intact. Bismarck's mistrust of democratic trends in Italy and the wish of the Italian monarchy to strengthen its position at home through close contact with the Dual Alliance, explain why the ideological motivation was mentioned in the official text. The preamble to the treaty explicitly states that the three monarchs wished 'to increase the guarantee of general peace, to strengthen the monarchical principle and to keep intact by that the social and political order of their states'. The Triple Alliance was made to defend the social and political 'status quo' in the three monarchies and perfectly reflected Germany's conservative foreign policy at this time.

The partners were forbidden to enter into alliances or agreements with any other partner (Art. I). In the case of France attacking Italy or Germany without provocation, the country that was not under attack was pledged to give military support (Art. II). The 'casus foederis' [a Latin expression that means "a case falling within the terms of a treaty"] was only to be invoked if one or two partners were attacked simultaneously by two or more Great Powers, without having been provoked by any of the partners in the Triple Alliance (Art. III). If any partner of the Triple Alliance should be forced into a preventive war, the other partners were pledged to remain neutral and were expected to give their partner military support at a later date (Art. IV). Any threat to security was to be met by agreeing to military cooperation, if this was thought to be necessary. Armistice and peace were only to be decided upon by all three partners and separate decisions were ruled out (Art. V).

Until now, German historians have always interpreted the Triple Alliance as being defensive. It may have been defensive at the beginning, but Article IV could be turned into an instrument of aggression: an 'unprovoked attack' on one of the partners of the Triple Alliance could always be camouflaged by an alleged threat from outside. The First World War, indeed, broke out in 1914 without a clear 'casus foederis' stated in the Triple Alliance. Austria-Hungary was not attacked by two or more Great Powers, and provocation came from Austria-Hungary. Neither Austria-Hungary nor Italy were obliged to assist Germany, if it chose to interpret the situation of July 1914 as one to justify a 'preventive war'. In this case, other partners could have remained neutral. Even Article I of the Dual Alliance did not cover the situation, because neither Austria-Hungary nor Germany were attacked in 1914 by Russia 'against the hope and the sincere wish of the two contracting parties'. Russia was provoked into mobilizing against the Dual Alliance by Germany and Austria-Hungary.

The direct consequence of these alliances was that after the Reich was founded in 1871, a new concentration of power in central Europe under German leadership came into being. When the Reich tried to make the Triple Alliance the basis for Germany's 'Weltpolitik', this action almost inevitably provoked an alliance between the Great Powers in eastern and western Europe.

The Triple Alliance Is Formed:
May 20, 1882

Bismarck's View of the Triple Alliance

by Otto von Bismarck

Otto von Bismarck was the main architect of the Triple Alliance formed on May 20, 1882. Bismarck served as chancellor of Germany from 1871 until 1890. A skilled diplomat, Bismarck charted the path for Germany's development as a strong European power after its unification in 1871. As a recently unified nation, Germany sought to protect its new status in Europe. Bismarck crafted foreign policy with this idea in mind. Alliances, with their prospects of mutual support and protection, appealed to Bismarck's view of diplomacy.

The following excerpt from Bismarck's memoirs conveys the chancellor's strategic thinking when the Dual and Triple Alliances were devised.

Bismarck also reflects on international policy in general. He writes that foreign affairs are fluid by nature, and that a change in atmosphere results in a change in policy. He is careful, therefore, not to endow the Triple Alliance with any sense of permanence.

T he triple alliance which I originally sought to conclude . . . was an alliance of the three Emperors with the further idea of bringing into it monarchical Italy. It was designed for the struggle which, as I feared, was before us; between the two European tendencies which [French emperor] Napoleon called Republican and Cossack, and which I, according to our present ideas, should designate on the one side as the system of order on

Otto von Bismarck, *Bismarck: The Man and the Statesman*. New York: Harper and Brothers, 1898.

a monarchical basis, and on the other as the social republic to the level of which the anti-monarchical development is wont to sink, either slowly or by leaps and bounds, until the conditions thus created become intolerable, and the disappointed populace are ready for a violent return to monarchical institutions in a Caesarean form. I consider that the task of escaping from this *circulus vitiosus* [vicious circle] or, if possible, of sparing the present generation and their children an entrance into it, ought to be more closely incumbent on the strong existing monarchies, those monarchies which still have a vigorous life, than any rivalry over the fragments of nations which people the Balkan peninsula. If the monarchical governments have no understanding of the necessity for holding together in the interests of political and social order, but make themselves subservient to the chauvinistic impulses of their subjects, I fear that the international revolutionary and social struggles which will have to be fought out will be all the more dangerous, and take such a form that the victory on the part of monarchical order will be more difficult. Since 1871 I have sought for the most certain assurance against those struggles in the alliance of the three Emperors, and also in the effort to impart to the monarchical principle in Italy a firm support in that alliance. I was not without hope of a lasting success when the meeting of the three Emperors took place at Berlin in September 1872, and this was followed by the visits of my Emperor to St. Petersburg in May, of the King of Italy to Berlin in September, and of the German Emperor to Vienna in the October of the next year. . . .

The Choice Between Austria and Russia

[Germany's lack of support from England] demanded an effort to limit the range of the possible anti-German coalition by means of treaty arrangements placing our relations with at least one of the Great Powers upon a firm footing. The choice could only lie between Austria and Russia, for the English constitution does not admit of alliances of assured permanence, and a union with Italy alone did not promise an adequate counterpoise to a coalition of the other three Great Powers, even supposing her future attitude and formation to be considered independently not only of French but also of Austrian influence. The area available for the formation of the coalition would therefore be narrowed till only the alternative remained which I have indicated.

In point of material force I held a union with Russia to have the advantage. I had also been used to regard it as safer, because I placed more reliance on traditional dynastic friendship, on community of conservative monarchical instincts, on the absence of indigenous political divisions, than on the fits and starts of public opinion among the Hungarian, Slav, and Catholic population of the monarchy of the Habsburgs. Complete reliance could be placed upon the durability of neither union, whether one estimated the strength of the dynastic bond with Russia, or of the German sympathies of the Hungarian populace. If the balance of opinion in Hungary were always determined by sober political calculation, this brave and independent people, isolated in the broad ocean of Slav populations, and comparatively insignificant in numbers, would remain constant to the conviction that its position can only be secured by the support of the German element in Austria and Germany. But the Kossuth episode, and the suppression in Hungary itself of the German elements that remained loyal to the Empire, with other symptoms showed that among Hungarian hussars and lawyers self-confidence is apt in critical moments to get the better of political calculation and self-control. Even in quiet times many a Magyar will get the gypsies to play to him the song, 'Der Deutsche ist ein Hundsfott' ('The German is a blackguard').

In the forecast of the future relations of Austria and Germany an essential element was the imperfect appreciation of political possibilities displayed by the German element in Austria, which has caused it to lose touch with the dynasty and forfeit the guidance which it had inherited from its historical development. Misgivings as to the future of an Austro-German confederation were also suggested by the religious question, by the remembered influence of the father confessors of the imperial family, by the anticipated possibility of renewed relations with France, on the basis of a *rapprochement* by that country to the Catholic Church, whenever such a change should have taken place in the character and principles of French statesmanship. How remote or how near such a change may be in France is quite beyond the scope of calculation. . . .

The Link with Austria

The traditions of international law from the time of the Holy Roman Empire, German by nation, and of the German confederation tended to the theory that between Germany as a whole and the Habsburg monarchy there existed a legal tie binding these central European territories together for purposes of mutual support. Practical effect had indeed rarely been given to this *consortium* in former ages; but it was possible to vindicate in Europe, and especially in Russia, the position that a permanent confederation of Austria and the modern German Empire was, from the point of view of international law, no new thing. These questions, whether the alliance would be popular in Germany, how far it could be justified by international law, were to me matters of subordinate importance, merely subsidiary to its eventual completion. In the foreground stood the question whether the execution of the design should be begun at once or deferred for a time, and with what degree of decision it would be advisable to combat the opposition which might be anticipated on the part of Emperor William—an opposition sure to be determined rather by his idiosyncrasy than by policy. So cogent seemed to me the considerations which in the political situation pointed us to an alliance with Austria that I would have striven to conclude one even in the face of a hostile public opinion. . . .

We must and can honourably maintain the alliance with the Austro-Hungarian monarchy; it corresponds to our interests, to the historical traditions of Germany, to the public opinion of our people. The influences and forces under and amid which the future policy of Vienna must be shaped are, however, more complex than with us, by reason of the manifold diversity of the nationalities, the divergence of their aspirations and activities, the influence of the clergy, and the temptations to which the Danubian countries are exposed in the Balkan and Black Sea latitudes.

We cannot abandon Austria, but neither can we lose sight of the possibility that the policy of Vienna may willy-nilly abandon us. The possibilities which in such a case remain open to us must be clearly realised and steadily borne in mind by German statesmen before the critical moment arrives, nor must their action be determined by prejudice or misunderstanding, but by an entirely dispassionate weighing of the national interests.

Relations with Russia

It has always been my endeavour to promote not merely the security of the country against Russian attacks, but also in Russia itself a peaceful tone, and a belief in the unaggressive character of our policy. Nor (thanks to the personal confidence which Czar Alexander III reposed in me) did I ever fail so long as I remained in office to turn the edge of the mistrust which again and again was aroused in his mind by misrepresentations on the part both of his own subjects and of foreigners, and occasionally by subterranean influences of a military kind from this side of the frontier.

At my first interview with him after his accession (in the Dantzic roads), and at all subsequent meetings, he was prevented neither by falsehoods disseminated in regard to the congress of Berlin, nor by the knowledge which he possessed of the Austrian treaty, from displaying towards me a good-will which at Skiernevice and at Berlin received authentic expression—a good-will which rested on personal trust in me. Even the affair of the forged letters placed in his hands at Copenhagen—an intrigue which by its shameless audacity was capable of producing the worst impression—was rendered innocuous by my mere disavowal. No less success had I at the meeting in October 1889 in dissipating the doubts which he had brought with him from Copenhagen, including the last, which concerned my own continuance in office. He was far better instructed than I when he put the question, whether I was quite sure of retaining my place under the new Emperor. I answered, as I then thought, that I was convinced that I possessed the confidence of Emperor William II, and did not believe that I should ever be dismissed against my will, because his Majesty, by reason of my prolonged experience in office, and the confidence which I had won for myself, not only in Germany, but in foreign Courts, had in my person a servant whom it was very difficult to replace. My assurance elicited from his Majesty an expression of great satisfaction, though he hardly seemed to share it unreservedly.

International policy is a fluid element which under certain conditions will solidify, but on a change of atmosphere reverts to its original diffuse condition. The clause *rebus sic stantibus* [meaning that a treaty can be terminated in extraordinary circumstances] is tacitly understood in all treaties that involve performance. The Triple Alliance is a strategic position, which in the face of the perils that were imminent at the time when it was concluded was politic, and, under the prevailing conditions, feasible. It has been from time to time prolonged, and may be yet further prolonged, but eternal duration is assured to no treaty between Great Powers; and it would be unwise to regard it as affording a permanently stable guarantee against all the possible contingencies which in the future may modify the political, material, and moral conditions under which it was brought into being. It has the significance of a strategic position adopted after strict scrutiny of political situation of Europe at the time when it was concluded, but it no more constitutes a foundation capable of offering perennial resistance to time and change than did many another alliance (triple or quadruple) of recent centuries, and in particular the Holy Alliance and the German Confederation. It does not dispense us from the attitude of *toujours en vedette* [the willingness to change].

A Uniform Standard of Time Is Established

by Stephen Kern

The formal worldwide decision to determine a universal standard of time occurred at the International Meridian Conference in October 1884. Delegates from across the globe met to designate a universal standard of time against which all time zones could be measured. Greenwich, England, which is at the zero longitude (meridian), was chosen to mark the beginning of every time zone in the world. Greenwich mean (or meridian) time, also called GMT, represents the time that it takes the earth to make a complete rotation, from noon to noon. Prior to the establishment of a uniform time standard, time was "local" and there existed no basis for synchronizing time zones.

This selection by Stephen Kern points out the importance of uniform, universal conceptions of time. The establishment of standard time affected the coordination of railroads, telegraph and other forms of communication, and military operations. The technological improvements and innovations of the era, then, complemented the use of standard time. Uniform time facilitated interaction among people, nations, and businesses separated by long distances. Kern also notes that some problems were incurred when countries made the conversion from local to universal time, and that it was several years before all nations adopted the GMT standard.

Stephen Kern is a professor of history at Ohio State University.

Stephen Kern, *The Culture of Time and Space, 1880–1918*. Cambridge, MA: Harvard University Press, 1983. Copyright © 1983 by the President and Fellows of Harvard College. All rights reserved. Reproduced by permission.

He has published five books and a number of articles on a variety of cultural and intellectual themes in modern European history.

T he most momentous development in the history of uniform, public time since the invention of the mechanical clock in the fourteenth century was the introduction of standard time at the end of the nineteenth century. A pioneer in promoting uniform time was the Canadian engineer Sanford Fleming, who in 1886 outlined some reasons for its adoption. The use of the telegraph "subjects the whole surface of the globe to the observation of civilized communities and leaves no interval of time between widely separated places proportionate to their distances apart." This system mixes up day and night as "noon, midnight, sunrise, sunset, are all observed at the same moment," and "Sunday actually commences in the middle of Saturday and lasts until the middle of Monday." A single event may take place in two different months or even in two different years. It was important to be able to determine local times and to know precisely when laws go into effect and insurance policies begin. The present system, he concluded, would lead to countless political, economic, scientific, and legal problems that only the adoption of a coordinated world network could prevent.

Adopting Universal Time

The most famous supporter of standard time, Count Helmuth von Moltke, in 1891 appealed to the German Parliament for its adoption. He pointed out that Germany had five different time zones, which would impede the coordination of military planning; in addition there were other time zones, he protested, that "we dread to meet at the French and Russian boundaries." When Fleming sent Moltke's speech to the editor of *The Empire* for publication, he did not dream that in 1914 the world would go to war according to mobilization timetables facilitated by standard time, which he thought would rather engender cooperation and peace.

Despite all the good scientific and military arguments for world time, it was the railroad companies and not the governments that were the first to institute it. Around 1870, if a traveler from Washington to San Francisco set his watch in every town he passed through, he would set it over two hundred times. The railroads attempted to deal with this problem by using a separate time for each region. Thus cities along the Pennsylvania Railroad were put on Philadelphia time, which ran five minutes behind New York time. However, in 1870 there were still about 80 different railroad times in the United States alone. The day the railroads imposed a uniform time, November 18, 1883, was called "the day of two noons," because at mid-day clocks had to be set back in the eastern part of each zone—one last necessary disruption to enable the railroads to end the confusion that had so complicated their functioning and cut into their profits. In 1884 representatives of twenty-five countries that convened at the Prime Meridian Conference in Washington proposed to establish Greenwich as the zero meridian, determined the exact length of the day, divided the earth into twenty-four time zones one hour apart, and fixed a precise beginning of the universal day. But the world was slow to adopt the system, for all its obvious practicality.

Problems with Time Conversion

Japan coordinated railroads and telegraphic services nine hours ahead of Greenwich in 1888. Belgium and Holland followed in 1892; Germany, Austria-Hungary, and Italy in 1893; but in 1899, when John Milne surveyed how countries throughout the world determined their time and its relation to Greenwich, there was still a great deal of confusion. Telegraph companies in China used a time that was approximately the same as in Shanghai; foreigners in coastal ports used their own local time taken from solar readings; and all other Chinese used sundials. In Russia there were odd local times such as that of St. Petersburg—two hours, one minute, and 18.7 seconds ahead of Greenwich. In India hundreds of local times were announced in towns by gongs, guns, and bells.

Among the countries in Western Europe, France had the most chaotic situation, with some regions having four different times, none of which had a simple conversion to Greenwich time. Each city had a local time taken from solar readings. About four minutes behind each local time was astronomical time taken from fixed stars. The railroads used Paris time, which was nine minutes and twenty-one seconds ahead of Greenwich. A law of 1891 made it the legal time of France, but the railroads actually ran five minutes behind it in order to give passengers extra time to board: thus the clocks inside railway stations were five minutes ahead of those on the tracks. In 1913 a French journalist, L. Houlle-vigue, explained this "retrograde practice" as a function of a national pride, expressed in the wording of a law of 1911 promoting the system that other countries of Europe had adopted twenty years earlier. The French law declared that "the legal time in France and Algeria is the mean Paris time slowed nine minutes and twenty-one seconds." Houllevigue pointed out the Anglophobic intent of the wording: "By a pardonable reticence, the law abstained from saying that the time so defined is that of Greenwich, and our self-respect can pretend that we have adopted the time of Argentan, which happens to lie almost exactly on the same meridian as the English observatory." In spite of their previous isolation the French finally took the lead in the movement for unified world time based on the guidelines of 1884. If the zero meridian was to be on English soil, at least the institution of world time would take place in France. So President Raymond Poincaré had Paris host the International Conference on Time in 1912, which provided for a uniform method of determining and maintaining accurate time signals and transmitting them around the world.

More Developments in Universal Time

The wireless telegraph made it all possible. As early as 1905 the United States Navy had sent time signals by wireless [telegraph] from Washington. The Eiffel Tower transmitted Paris time in 1910 even before it was legally declared the time of France. By 1912 the system was expanded with installations in Nancy, Charleville, and Langres so that the entire country could receive the same signals simultaneously. Houllevigue boasted that Paris, "supplanted by Greenwich as the origin of the meridians, was proclaimed the initial time center, the watch of the universe." The observatory at Paris would take astronomical readings and send them to the Eiffel Tower, which would relay them to eight stations spaced over the globe. At 10 o'clock on the morning of July 1, 1913, the Eiffel Tower sent the first time signal transmitted around the world. The independence of local times began to collapse once the framework of a global electronic network was established. Whatever charm local time might have once had, the world was fated to wake up with buzzers and bells triggered by impulses that traveled around the world with the speed of light. . . .

The proponents of world time were few, and none of them

(aside from Moltke) were well known beyond the narrow circle of fellow reformers. Nevertheless the concept of public time was widely accepted as a proper marker of duration and succession. There were no elaborate arguments on its behalf because there seemed to be no need. The passion in the debate about homogeneous versus heterogeneous time was generated rather by those novelists, psychologists, physicists, and sociologists who examined the way individuals create as many different times as there are life styles, reference systems, and social forms.

3 The International Meridian Conference Sets
Greenwich Mean Time: October 13–22, 1884

The Importance of Greenwich Mean Time

by the International Meridian Conference

As a result of the International Meridian Conference held October 13 to 22, 1884, uniform time was established throughout the world. This international conference, convened by U.S. president Chester A. Arthur, took place in Washington, D.C. Twenty-five nations were represented by forty-one delegates, who discussed how to coordinate longitudinal measures to a single standard of time.

The following excerpt is from the conference's final act and contains both the resolutions and the votes that accompanied them. Each resolution underscores the technical ramifications of setting a standard for universal time, such as whether the initial meridian should begin at the Greenwich Observatory in England and whether there should be a universal day for beginning the uniform standard. Other principles enacted at the International Meridian Conference included the following: Midnight would be the start of astronomical and nautical days everywhere; the universal day would begin at the Greenwich Observatory at midnight and be counted on a twenty-four-hour clock; and technical studies about applying the decimal system to the division of time and space would be supported. Although the resolutions were adopted by the conference as a whole, some nations refrained from immediate cooperation. France, disappointed by the British-American domination of the proceedings, did

International Meridian Conference, "Final Act: Resolutions and Voting," Washington, DC, October 1884.

not adhere to universal standard time until 1911. Nevertheless, the overall conclusion of the International Meridian Conference was that adopting a single meridian as a time standard was more desirable than having numerous local times.

C ount Lewenhaupt, Delegate for Sweden, . . . proposed that the resolutions passed by the Conference should be formally recorded in a Final Act, stating the votes on each resolution that was adopted.

The Conference took a recess, in order to allow the Delegates to examine a draft of the Final Act.

After the recess the Final Act was unanimously adopted, as follows:

FINAL ACT.

The President of the United States of America, in pursuance of a special provision of Congress, having extended to the Governments of all nations in diplomatic relations with his own, an invitation to send Delegates to meet Delegates from the United States in the city of Washington on the first of October, 1884, for the purpose of discussing, and, if possible, fixing upon a meridian proper to be employed as a common zero of longitude and standard of time-reckoning throughout the whole world, this International Meridian Conference assembled at the time and place designated; and, after careful and patient discussion, has passed the following resolutions:

I.

"That it is the opinion of this Congress that it is desirable to adopt a single prime meridian for all nations, in place of the multiplicity of initial meridians which now exist."

This resolution was unanimously adopted.

II.

"That the Conference proposes to the Governments here represented the adoption of the meridian passing through the centre of the transit instrument at the Observatory of Greenwich as the initial meridian for longitude."

The above resolution was adopted by the following vote:

In the affirmative:

Austria-Hungary,
Chile,
Colombia,
Costa Rica,
Germany,
Great Britain,
Guatemala,
Hawaii,
Italy,
Japan,
Liberia,

Mexico,
Netherlands,
Paraguay,
Russia,
Salvador,
Spain,
Sweden,
Switzerland,
Turkey,
United States,
Venezuela.

In the negative:

San Domingo.

Abstaining from voting:

Brazil,   France.

Ayes, 22; noes, 1; abstaining, 2.

III.

"That from this meridian longitude shall be counted in two directions up to 180 degrees, east longitude being plus and west longitude minus."
This resolution was adopted by the following vote:
In the affirmative:

Chile,
Colombia,
Costa Rica,
Great Britain,
Guatemala,
Hawaii,
Japan,

Liberia,
Mexico,
Paraguay,
Russia,
Salvador,
United States,
Venezuela.

In the negative:

Italy,
Netherlands,
Spain,

Sweden,
Switzerland.

Abstaining from voting:

Austria-Hungary,   Germany,

Brazil,	San Domingo,
France,	Turkey.

Ayes, 14; noes, 5; abstaining, 6.

IV.

"That the Conference proposes the adoption of a universal day for all purposes for which it may be found convenient and which shall not interfere with the use of local or standard time where desirable"

This resolution was adopted by the following vote:

In be affirmative:

Austria-Hungary,	Mexico,
Brazil,	Netherlands,
Chile,	Paraguay,
Colombia,	Russia,
Costa Rica,	Salvador,
France,	Spain,
Great Britain,	Sweden,
Guatemala,	Switzerland,
Hawaii,	Turkey,
Italy,	United States,
Japan,	Venezuela.
Liberia,	

Abstaining from voting:

Germany,	San Domingo.

Ayes, 23; abstaining, 2.

V.

"That this universal day is to be a mean solar day is to begin for all the world at the moment of mean midnight of the initial meridian, coinciding with the beginning of the civil day and date of that meridian and is to be counted from zero up to twenty-four hours."

This resolution was adopted by the following vote:

In the affirmative:

Brazil,	Liberia,
Chile,	Mexico,
Colombia,	Paraguay,
Costa Rica,	Russia,

Great Britain,
Guatemala,
Hawaii,
Japan,

Turkey,
United States,
Venezuela.

In the negative:

Austria-Hungary,   Spain.

Abstaining from voting:

France,
Germany,
Italy,
Netherlands,

San Domingo,
Sweden,
Switzerland.

Ayes, 15; noes, 2; abstaining, 7.

VI.

"That the Conference expresses the hope that as soon as may be practicable the astronomical and nautical days will be arranged everywhere beginning at midnight."

This resolution was carried without division.

VII.

"That the Conference expresses the hope that the technical studies designed to regulate and extend the application of the decimal system to the division of angular space and of time shall be resumed, so as to permit the extension of this application to all cases in which it presents real advantages."

The motion was adopted by thc following vote:

In the affirmative:

Austria-Hungary,
Brazil,
Chile,
Colombia,
Costa Rica,
France,
Great Britain,
Hawaii,
Italy,
Japan,
Liberia,

Mexico,
Netherlands,
Paraguay,
Russia,
San Domingo,
Spain,
Switzerland,
Turkey,
United States,
Venezuela.

Abstaining from voting:

Germany,   Sweden.
Guatemala,

Ayes, 21; abstaining, 3.
Done at Washington the 22nd of October, 1884.
C.R.P. RODGERS
President.
R. Steachey, J. Jannsen, L. Cruls,
Secretaries.
The following resolution was then adopted unanimously
"That a copy of resolutions passed by this Conference shall be communicated to the Government of the United States of America, at whose instance and within whose territory the Conference has been convened."

Mr. Rutherford, Delegate of the United States, then presented the following resolution:

"Resolved, That the Conference adjourn, to meet upon its call of the President, for the purpose of verifying the protocols."

This resolution was then unanimously carried, and the Conference adjourned at half past three, to meet upon the call of the President.

The Berlin Conference Divides Africa:
February 26, 1885

Africa Becomes a Colonial Power

by H.L. Wesseling

Late-nineteenth-century Europe was characterized partly by imperialism and colonialism. Imperialism is the exercise of power by a state beyond its borders. Colonialism, which is a form of imperialism, is when a state sets up an administrative system to control another group of people. European nations such as Britain, France, Germany, Belgium, and Portugal sought to colonize Africa, and by the 1880s nations were scrambling to set up formal empires on the African continent. This prompted Otto von Bismarck, chancellor of Germany and master diplomat, to call a conference as a means of addressing the competition among imperialists' ventures in Africa. The Berlin Conference, which ran from November 15, 1884, to February 27, 1885, established guidelines for the partitioning of Africa, which was regarded as an international no-man's-land. African rulers were excluded from the Berlin Conference.

H.L. Wesseling's article discusses the ramifications of the Berlin Conference. He writes that diplomatic interests were most important during the European congress. Imperial interests were the second-most important factor, with African interests being subordinated to both. Wesseling also notes that most of the procedures devised at the conference were not consistently upheld, and no formal territorial boundaries were established. Nonetheless, the conference held symbolic significance for Europe's colonialist expansion. According to Wesseling, rival interests in Africa constituted a "game" for the Europeans.

Wesseling states that the partition of Africa was a turning point and a prelude to more direct colonial rule. He concludes that the Berlin Conference was part of a larger process of European colonization—and eventual decolonization—that drew Africa into global politics and economics.

H.L. Wesseling is professor emeritus of contemporary history at Leiden University in the Netherlands, where he taught for many years. He has published extensively on the subject of European expansion.

P aradoxes and problems, the Berlin Conference offers us plenty of them. In the first place there is the strange contrast between what one originally expects to find as the subject of discussions at Berlin and what was discussed in actual fact. The words 'Berlin Conference' have historical connotations. They suggest something in the tradition of the great diplomatic congresses of the nineteenth century, comparable to the congresses of Vienna, Paris, Berlin 1878. The very name Berlin conjures up the idea that something important happened.

The Conference in Context

Of course it was only due to some coincidence that the Conference took place on the banks of the Spree and not of the Seine. Or rather it was due to French prudence that suggested keeping a rather low profile at what was supposed to become an anti-British manifestation. All the same, the place of the Conference, Berlin, seems to have a certain symbolic meaning. Here we are in the centre of power politics and world diplomacy at that time. President of the Conference is none other than [Otto von] Bismarck himself, *Realpolitiker par excellence*, master of diplomacy, connoisseur of power politics, architect of Germany's imperial power, father of coalitions, partitions, and wars. [A *Realpolitiker* takes an approach to statesmanship that is realistic and opportunistic rather than moralistic.]

But as soon as we start reading the Berlin Act and Protocols we know that we are in a very different world. This is not a conference like the Congress of 1878 or the Paris Conference of 1856. This is not the past but the future. It reminds one of the Peace Conference of The Hague, or the League of Nations or the United Nations. The political discourse is not that of power politics but of international law. The pros and cons of every word and sentence are discussed. The delegates carp at everything. Paragraphs are written and rewritten a dozen times. It is like the drafting of the resolutions of the United Nations. . . . One is reminded not of Otto von Bismarck but of Woodrow Wilson. To put it very briefly, it is immediately obvious that something very unimportant and something very unreal is going on. What makes it even more unreal, of course, is that we all know that if ever there was a period of power politics this was it. This period was to end with the great war of 1914. This Conference took place while a major international crisis was going on, the Anglo-Russian conflict in Afghanistan that was to come to a climax only a few weeks after the Conference and threatened to develop into the greatest war since the Crimean. The tragedy of Gordon at Khartoum had just taken place. [British general John Gordon was killed by natives in Sudan.] For Britain, it was a desperate situation. As [British statesman] Lord Milner wrote: 'Everything, yes absolutely everything, seemed bent upon going wrong at one and the same time.' Or in the words of [George] Granville, the Foreign Secretary, it was 'dreadful, jumping from one nightmare into another'. The French government also had serious problems. There was a stalemate in Madagascar and a difficult war was going on with China. The defeat of Langson was about to come. [During the Sino-French War, the Chinese defeated the French at Langson.] A few months later both the [Prime Minister William] Gladstone and the [Jules] Ferry cabinets would fall. Yet here we see these people seriously, slowly, and peacefully discussing matters of international law, codes of conduct, navigation acts, and postal unions. One should not be surprised that Bismarck lost all interest in the Conference, right from the start. His only role was to open and close it. Apart from that he was not to be seen, an intention which he made clear in the opening session.

Of course, another Berlin Conference was going on at the same time, not in the conference room but in the lobby. And here a different language was spoken—that of *Realpolitik*—and matters of practical importance were discussed. The Free State got its international recognition and its boundaries were fixed, albeit in a somewhat provisional way. When we discuss the meaning of the Berlin Conference, this side of it should, of course, be taken into account as well. But let us first return to the Conference itself and have a closer look at the diplomacy.

All the participating powers came to the Conference not only with aims concerning the official points of discussion as mentioned in the agenda, but also with considerations of wider imperial interest and of international politics in general. These considerations formed, as it were, three concentric circles of three different layers of interests, in order of importance. The African interest was subordinated to the imperial interest and that again to general diplomatic interests. This hierarchy of values is very clearly illustrated by the Dutch diplomacy at the Conference, a subject not further discussed here but nevertheless not wholly uninteresting. The first layer was that of the Congo problem itself. The Dutch had very considerable trading interests there. Dutch businessmen were organized in a powerful lobby. But the Foreign Minister told his representative at Berlin in no uncertain terms that these considerations were of less importance than the others, the imperial interests and the international situation. A major aim of Dutch diplomacy was to restrict the working of the Conference strictly to Africa. The great fear was that the decisions—for example those about effective occupation—would be given a more general significance and be considered as applicable to the Netherlands Indies as well. Restriction of the work of the Conference to Africa was therefore an important priority. But this, the imperial interest, was again subjected to the overruling priority of Dutch diplomacy, the maintenance of good relations with its two powerful neighbours: Germany in Europe and Britain in Asia. The Dutch position could be summarized as follows: one talked about Africa but one thought about Indonesia and was concerned about Europe.

These three levels of interest are to be found with each power that had an interest at stake in the Conference. (There were of course nations represented in whose case it is difficult to find such interests at all, for example Sweden or Russia). For Britain there was the Congo itself—the British had a trading interest there. But more important was the Niger which was to be protected as a British sphere. Even more important were the imperial problems: Egypt, the Suez Canal, and so on. Above all there was the international situation at large, the danger of a united German-French front, of a none too splendid isolation at a moment of deteriorating relations with Russia.

For France there was the Congo itself—[Pierre] Brazza's [founder of a French colony in West Africa] acquisitions had become a subject of national pride and grandeur. The greater interest of imperial strategy was of course dominated by France's desire to reopen the Egyptian question with the help of Germany. Finally there were considerations of international relations, the dilemma between continental and overseas priorities, the choice between Alsace-Lorraine and the colonies, or, as the popular poet and nationalist Paul Déroulède put it, between 'two lost sisters and twenty servants'.

For Germany of course the priorities were even more obvious: even in his 'colonial year' (1884) Bismarck did not for a single moment forget that, as he said, his 'map of Africa was in Europe', that the empire would never be more than of very marginal interest to Germany. His colonial conversion has been discussed a hundred times. In essence the explanation seems to be quite simple and was perfectly formulated by Bismarck himself in a speech to the Reichstag of 26 January 1899, when he said of the German colonists: 'They cannot prove that it is useful for the Reich. I, however, cannot prove that it is harmful to it, either.' In such a situation the policy to follow is not difficult: grab what one can get at the lowest possible price and as long as it does not harm major political interests.

Thus we see with all the powers that general strategic considerations rather than African interests determined their African policy. . . . At the level of the official mind one sees the weighing of various interests. In this process the general political interests of the nation as seen by the decision-makers are what counts. Perhaps it is not a surprising conclusion, because it is, after all, what Foreign Secretaries and Cabinet Ministers are supposed to do: formulate the national interest and act accordingly. But it explains to a certain extent one of the most remarkable aspects of the Berlin Conference, that it seems to have been one of those rare things in world history, a competition with only winners, a lottery without blanks. Germany was satisfied. It had an interest in free trade in the Congo and that was guaranteed. It wanted to be taken seriously as a colonial power. That had been the cause of the friction with England in 1884 when the British had originally refused to do so. Now it got Britain's recognition. And even more: it received Gladstone's personal blessing. On 12 March 1885 he said in the Commons: 'If Germany is to become a colonising power, all I say is 'God speed her!' She becomes our ally and partner in the execution of the great purposes of Providence for the advantage of mankind.' Bismarck was not very impressed: he liked Gladstone about as much as [German chancellor] Helmut Schmidt admired [U.S. president] Jimmy Carter and must have fully agreed with Queen Victoria's judgement that he was 'an old, wild and incomprehensible man'.

For France the Conference was a success. Brazza's new colonial acquisitions—source of great French pride—had been secured. The pre-emptive right on the Free State's territories opened more possibilities for the future. Its nuisance value *vis-à-vis* Britain had been demonstrated. Its relation with Germany was improved, not fundamentally, but at least as far as possible.

Britain of course was the country against which the Conference was originally arranged. But it did very well and survived without great problems. Britain had nothing against free trade in the Congo. It was much more interested in the Niger and it had succeeded right from the beginning in obtaining an agreement that this river was not to be the subject of internationalization but considered part of the British sphere of interest. It had fought hard on the issue of effective occupation and the protectorate system. This issue was explicitly meant to be turned into a weapon against the supposed pretentions of a British Monroe doctrine for Africa. Even this fight it had won. And it had demonstrated that in so far as Africa was concerned, Germany had many more interests in common with Britain than with France. An anti-British Continental League had become very improbable.

The Conference was perhaps the greatest triumph for a state that was not even officially present there, the Congo Free State. It got its recognition and reached agreements on its boundaries with France and Portugal—as well as territorial recognition by Germany. Its existence was welcomed at the final session as a great step in the history of human civilization.

There *was* of course a loser at the Conference, namely Portugal. But there was nothing unusual in that. It would be hard to find any colonial success for Portugal in the entire nineteenth or for that matter even the eighteenth century. The history of Portuguese colonization since the seventeenth century is essentially a history of decline.

Success for the European Powers

How was this possible? How was a conference possible that was a success for practically everybody? There are various reasons. In the first place much of what was considered to be a success

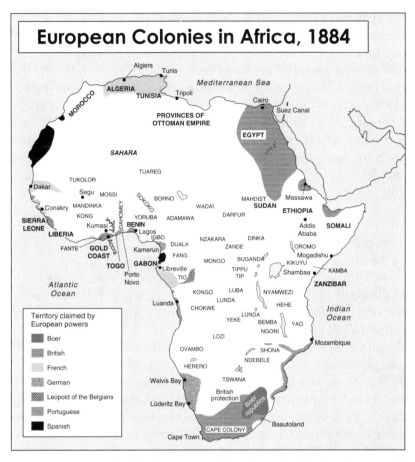

was a success only in the supposition that the Berlin Act would be applied in actual practice. As is well known, this was not to be the case. Many of the arrangements were to be violated, particularly those concerning free trade and freedom of navigation. The Dutch traders, for example, were soon to find out that what they had welcomed and praised as a triumph for free trade was, in fact, a defeat of their trading interests. After that, they were no longer all that happy with the Berlin Conference. The second reason was that the various nations approached the African problems in different terms and therefore had different definitions of what was a success. Where the French were, generally speaking, more interested in political concepts, the British approach was a more commercial one. This made it possible to make bargains with both parties considered to be successes. This was the case not only in Berlin, but also with regard to later agreements, for example, the one between France and Britain on West Africa in 1898.

But there was another and more fundamental factor at work, one that was characteristic of the partition of Africa in general, namely the simple fact that there was indeed something for everybody. This is perhaps the most curious and unique aspect of the partition and of this stage of imperialism in general. There were, of course, fights and negotiations, bargains and deals. This was necessary if only for domestic reasons: one could not give away the national interest without a fight. But when all was said and done, there really was something for everybody. [Belgian king] Leopold's comparison of Africa with a 'magnificent cake' was a very appropriate one, because it illustrated two important aspects of it at the same time. In the first place, there was no serious problem as long as there was a piece of cake for everybody. In the second place, the fight was about a cake, that is to say, about a luxury, not a vital need. . . . This was an essential element of the Berlin Conference and of partition in general: these were not vital issues. Africa was not the Balkans, let alone Alsace-Lorraine. In his memoirs Bismarck devotes a full chapter to the Berlin Congress of 1878. He does not even mention the Conference of 1884.

Thus there was satisfaction for everybody, but was there really reason to be so satisfied? Here again we enter a world of paradoxes. Most of the decisions that were taken were not respected and the one proposal that was rejected was to have a certain impact after all. The Conference decided to guarantee freedom of trade in an immense region, the so-called Conventional Congo Basin. This was never implemented. Leopold promised to create a state without customs. The Congo State, however, was to become a state of monopolies and exclusive rights under the domanial system. There was to be freedom of navigation on the Congo and Niger rivers. This was never realized, neither on the Congo nor on the Niger. . . . There were long debates and high-spirited speeches about humanitarian ideals. The results, however, were disappointing. Maybe the image of the Free State is too much dominated by pictures of amputated hands and feet. But it could hardly be considered as a triumph of humanitarian ideals either. And then, of course, there was the famous fight about effective occupation. Here the text was so much watered down, by restricting it to new requisitions—thus not to old ones—to the *coasts* only—thus not to the interior—and to *oc-*

cupations—thus not to protectorates—that in fact it became meaningless. But the ultimate paradox is that although the principle was rejected, it nevertheless acquired some influence, because the principle of effective occupation was to gain a certain exemplary value, a normative meaning. The British victory on this issue was very much a Pyrrhic victory.

The Significance of the Conference

What then was the significance of the Berlin Conference? What role did it play in the partition? This is a question to which many answers have been given, of which I shall discuss only a few.

The first answer is a simple one: the Berlin Conference partitioned Africa; it drew the boundaries of the various European possessions. As Kwame Nkrumah put it: 'the original carve-up of Africa was arranged at the Berlin Conference of 1884. It is a theory that can still be heard today, even from serious authors like Basil Davidson. We might call this the 'myth of the Berlin Conference' because it is, of course, not true. The Conference itself drew only one boundary, that of the free trading zone. Some of the bilateral treaties between the powers and the Free State were signed during, though not actually at, the Conference. But there was only one—that with Germany, which had already been signed before the Conference—that was accompanied by a map indicating the territory of the Free State. The French and Portuguese treaties describe the boundaries between their possessions and those of the Free State. There was no 'carve-up' at the Berlin Conference.

The second theory is that of the Hinterland, a story still to be found in many textbooks, for example in the otherwise excellent book by R.R. Palmer *A History of the Modern World.* Palmer writes that 'a European power with holdings on the coasts had prior rights in the back country'. Nothing of this is to be found in the Berlin Act, but again it is a very persistent myth. . . . The first mention of it [is found] in a French textbook of 1918. But there are even older ones. The original source of it might be a popular German textbook of 1907, D. Schäfer's *Weltgeschichte der Neuzeit.* This exercise in historiography, interesting though it is, should not occupy us here.

The next theory is that the Berlin Conference laid down the 'ground rules' for the scramble or formulated the 'code of conduct' for the partition. It is true that the Berlin Act laid down a few general principles: two articles out of the thirty-eight of the General Act are devoted to them. These articles, however, are not directives for partition, but diplomatic precautions to avoid international problems. Another very popular metaphor is that the Berlin Conference 'fired the starting gun for the partition'. These kinds of literary formulations are always rather flexible, but if this was a start it was a false one because most of the runners were already well under way. If the Conference tried to do anything it was rather to call back the competitors. It was, to use W.J. Mommsen's phrase, a 'holding operation'. But it was a holding operation that failed. . . .

What then is the meaning of the events of 1884–5? In retrospect they seem very much to be a turning point, a watershed in the history of British imperialism and that of the partition of Africa. They mark the end of an era that could be labelled 'the post-Napoleonic era'. It is amazing how much the history of Britain's paramountcy in the nineteenth century reminds one of the US in the twentieth. Both were based on economic superiority. Both supremacies only became manifest after wars to which, originally, they had not been a party. Both used the instruments of naval power and informal empire. Both preferred informal to formal empire. Both originally profited from the absence of serious competition: before 1870 Germany did not exist, nor did Italy. Austria and Russia were essentially continental powers and they were haunted by domestic problems. France had no naval power of any importance. As for the US, after 1945 Japan and Germany were defeated, Russia was victorious but badly wounded, Europe was a shambles. By 1980 the picture was very different for the US: the Soviet Union had become a major power, Europe was restored, Japan had become more of a rival than it had ever been. The same was true of Britain in the 1880s: the US, Japan, and especially Germany had become economic rivals, Russia was a danger. In 1885 a naval scare went through England. The very basis of its empire was threatened. Finally British imperialism was to be badly damaged by a war that was more of a domestic and political crisis than a military one. In this respect the Vietnam war can be compared to the South African war of 1899–1902. In 1914 the bell tolled for the British Empire. But in 1885 the writing was already on the wall.

1885 was also a turning point in the history of the partition of Africa. Here of course we enter a great debate, the one about the causes and the chronology of the scramble. This debate essentially comes down to two questions: why did the scramble take place when it did, and in the way it did? But there is a preliminary question: when did the scramble take place? Here the discussion is about the beginning. That the scramble was over by about 1912 is no matter for discussion. But its beginning is a far more complicated issue. Historians have taken different positions on this. Many years and events have been suggested: 1884, because of the Anglo-Portuguese treaty; the ratification of the Brazza-Makoko treaties in November 1882; the British occupation of Egypt earlier that year; the French occupation of Tunisia in 1881 or the forward policy introduced in French West Africa in 1879. The importance of this debate does not of course lie in the finding of the exact date but in the fact that every date implies a certain theory about the causes of or the responsibility for the scramble. . . .

A Stable Period

The partition of Africa was in the 'best' tradition of European politics: it was about territorialization, about borders and boundaries. We often speak of the 'artifical boundaries of Africa', but were they any more artificial than the European ones? In a way the partition was nothing but the entire history of Europe since the Middle Ages all over again, but in a condensed form: 400 years of history repeated within 30 years! But there was one great difference: in European history annexations and wars were followed by peace treaties, boundaries, and maps. In Africa, they started with maps and treaties and war came later, if at all. And if war came, it was not among Europeans but between Europeans and Africans. This explains one of the most curious phenomena of the partition, its peacefulness. Most of the partition took place between 1885 and 1895. When one looks at the map of 1895 in [J.S.] Keltie's *Partition of Africa* one sees that by then the partition was nearly complete. On the other hand, we know that the decade 1885–95 was the most peaceful in modern history. . . . In that period there was only one great war (between China and Japan) and one smaller war (between the Congo State and the Arabic slave traders). This is to say that during the partition itself there were practically no European or colonial wars.

There are two possible explanations for this strange phenomenon. In the first place there is the danger of a conceptual fallacy:

maybe the application of violence as used in Africa did not fit with the criteria of traditional war, was therefore not classified as such and thus not counted. But there is another explanation as well, namely that during the heyday of partition very little actually happened in Africa. What these maps illustrate is not reality but fiction. They illustrate the agreements on boundaries as reached in European chanceries and offices, not the occupation itself. This came later and cannot be dated so easily.

This order of things was very different from European history. It was not so much European history repeated as European history upside down. In European history there is first annexation, war, and so on and finally there are maps that represent the result of all this. In Africa the maps came first, maps that in the beginning represented nothing but themselves. Normally a map is a representation of reality in a realistic or coded form. Not so 'The Map of Africa by Treaty'. Here there was no reality to be represented. Here, to use a well-known expression from the 1960s, 'the medium was the message'.

This explains much of the peacefulness of the 1885–95 decade. Not much happened except on paper. It also explains why the Europeans could so easily avoid getting involved in major conflicts. Territorial questions were settled in advance. Moreover, these were arrangements about regions that those involved did not know and certainly did not care much about. All this lightheartedness is perfectly illustrated in a speech by [Prime Minister] Lord Salisbury in the [House of] Lords in 1890, where he said: 'I will not dwell upon the respective advantages of places which are utterly unknown not only to your Lordships, but to the rest of the white human race. Africa was 'very light soil' indeed! The rivalry between Russia and Britain on the North West Frontier was known as the Great Game. But compared to Africa, this was not a game, but business. In Africa—apart from the Mediterranean—European rivalry never became more than a game. . . .

The Partition of Africa in World History

Politically speaking the partition was a period of transformation. It was the prelude to colonial rule. Full colonialism came to Africa in about 1914. By that time Africa was almost entirely under colonial rule. But the colonial period was a very short one—particularly in the time perspective of a continent where the cradle of humanity stood. It lasted only half a century. By the end of the 1960s it was virtually over. Short as it may have been, however, it was a very painful period in African history. The loss of sovereignty and dignity, the subordination to the rule of 'alien races' as our nineteenth century colleagues used to say, were a sad page in the book of African history. In this respect decolonization was a major change.

Economically speaking however this period does not seem to be such an important phase in European colonialism. As seen from the post-colonial perspective the period was in fact only a rather unimportant stage in a much longer process. This process brought about the incorporation of Africa into the world economy and the spread of industrial civilization over the continent. From this point of view the partition was not a major episode. The major events came later. The transfer from commercial to economic exploitation took place in the 1920s and these years were more of a watershed than the 1880s. On the other hand decolonization did not bring an end to this process. On the contrary, dependency and interconnection became all the greater.

As seen from the perspective of world history the partition of Africa was just a chapter of a much bigger story, that of the partition of the world, and the late nineteenth century only an episode in a long-term process that lasted for a few centuries. This process could simply be called the division of the world or the expansion of Europe and it started long ago. The first division of the world took place in about 1500. In 1492 Columbus discovered America and in 1498 Vasco da Gama rounded the Cape of Good Hope. 'This', as [French historian] Fernand Braudel said, 'forced on Europe an extremely grave choice: either to make use of Christopher Columbus' discovery and opt for America, or to exploit the discovery of the continuous sea links round the Cape of Good Hope to its limits and batten on to Asia.' Europe did both and even at the same time, but the results were quite different. In the rather empty world of the Americas it founded New Spain, New England, New France, and so on. In short, it created a new Europe, Europe overseas, better known as the 'New World'. This was Europe's long-standing achievement. Perhaps it was the greatest creative enterprise in the entire history of Europe. In Asia something very different happened. Here the Europeans faced densely populated regions, highly centralized polities, complicated and sophisticated trading networks. Europe joined the game of Asian trade but it never took over. It tapped the riches of Asia but it did not change the fabric of Asian society—at least not in this first stage, before the industrial revolution.

This then was the first division of the world. Next to the old world of Europe, Africa, and Asia, there was now a New World. But this division did not last out when, after the eighteenth century, the world was going through the greatest transformation it has known, the industrial revolution. This created a new dichotomy and a new world order. A part of the old world, Europe, abandoned the old pattern of life. In this it was followed by the New World. Europe and America joined to become the developed world. Asia and Africa were left as the world of the old order. This new dichotomy—and this was the important thing— was hierarchical, one between superior and inferior. Europe was able to exploit Asia, which had not been the case before.

In both periods Africa was touched very little by European expansion. During the first period, it was mostly uninhabitable for Europeans and thus offered no opportunity for creating a new world as in America. It did not have a large trading network for Europeans to join nor ready-made riches to exploit, as did Asia. Its major commodity was manpower, that is to say slaves. In the second, the industrial period, it did not offer many profitable opportunities for investments in agriculture or industry, nor was it a market comparable to those of Asia. Libya is about as big as the Dutch East Indies, the Anglo-Egyptian Sudan only some 10 per cent smaller than India. All the same nobody will argue that they were of the same interest for European colonialists. Up to the First World War, apart from Algeria and South Africa, no colony in Africa was comparable to the European possessions in Asia. There might have been imperialism in Africa. The expansion of Europe took place in America and, although to a lesser extent, in Asia.

What, then, should be our conclusion on European colonialism and European expansion? As for colonialism, the same verdict seems to be applicable as [French political philosopher] Montesquieu pronounced on the history of slavery: 'an institution implying misery and destitution for great numbers while it made life easier for just a few but eventually was of real use to nobody at all'.

Louis Pasteur's Rabies Vaccine Is Successful:
July 6, 1885

The Antirabies Vaccine Was Pasteur's Greatest Triumph

by René Dubos

In the nineteenth century rabies was one of the deadliest diseases for animals and humans. This disease, which is almost always fatal, can be transmitted to humans through the bite of an infected animal. Symptoms of the illness are progressive and include headache, fever, anxiety, confusion, delirium, hallucinations, and insomnia.

Louis Pasteur (1822–1895), a renowned chemist and biologist, worked for five years to isolate and cultivate the rabies virus. After successfully cultivating the virus, Pasteur was able to create a vaccine. The treatment was successful on animals.

In the following selection René Dubos comments that Pasteur was reluctant to test this treatment on humans. Dubos relates the story of nine-year-old Joseph Meister, who had fourteen bites from a rabid dog. On July 6, 1885, the rabies-infected and seemingly doomed boy was brought to Pasteur and inoculated repeatedly over the next twelve days. The treatment was a success, and Meister returned home in good health.

Dubos maintains that the antirabies vaccine was Pasteur's greatest triumph. This achievement led to the development of immunology and vaccine techniques for various viral diseases. In 1888, as a result of the scientist's success, the Pasteur Institute opened in

René Dubos, *Pasteur and Modern Science*, edited by Thomas D. Brock. Madison, WI: Science Tech, 1988.

France with the goal of furthering research in biology. Dubos cred-
its Pasteur's experimentation boldness and mental courage with
helping the scientist reach new heights of discovery.

René Dubos was a microbiologist and an experimental patholo-
gist. He was both a researcher and a prolific, Pulitzer Prize–winning
author. In 1980 the René Dubos Center for Human Environment
opened in his honor. He was professor emeritus at Rockefeller
University.

I t is . . . the antirabies treatment which is usually cited as Pasteur's greatest triumph and claim to immortality, and which
established microbiological sciences in the popular mind and
in the practice of medicine. Rabies had long had a firm hold on
public imagination and was the epitome of terror and mystery. It
was therefore well suited to satisfy Pasteur's longing for romantic [idealized, heroic, adventurous] problems. It combined a
supreme challenge to the experimenter and his method, and the
chance to capture the interest of the medical and lay public by a
spectacular achievement. In fact, Pasteur was right in the selection of this seemingly hopeless problem. The Pouilly le Fort experiment on anthrax [where Pasteur tested the anthrax vaccine]
had rendered the public conversant with the doctrine of immunization, but it was the prophylaxis of rabies that made of microbiological science an established religion and surrounded its cre-
ator with the halo of sainthood.

From Animals to Humans

In the first phase of the rabies work, Pasteur showed that the
spinal cords of rabbits dead of the disease could be rendered almost nonvirulent by keeping them for two weeks in sterile dried
air. Specifically, the technique consisted in keeping the spinal
cord in a container with caustic potash to prevent putrefaction,
and allowing penetration of oxygen to attenuate the virus. The
famous portrait painted by [Finnish painter Albert] Edelfeldt
shows Pasteur absorbed in the contemplation of one of these
flasks. By inoculating dogs with emulsions of progressively less
attenuated cord, it was possible to protect the animal against inoculation with the most virulent form of virus. Under normal
conditions of exposure rabies develops slowly in humans as well
as in animals. For example, a person bitten by a mad dog ordi-

narily does not display symptoms of the disease until a month or more after the bite. This period of incubation therefore appeared long enough to suggest the possibility of establishing resistance by vaccinating even after the bite had been inflicted. Experiments made on dogs bitten by rabid animals, and then treated with the vaccine, gave promising results. Would the same method be applicable to human beings bitten by rabid animals and still in the incubation period of the disease?

The story of the mental anguish Pasteur experienced before daring to proceed from animal experiments to the treatment of human disease has often been told. The decision to apply rabies vaccination to the human was forced upon him when a young boy, Joseph Meister, was brought from Alsace for treatment on July 6, 1885, suffering from rabid dog bites on the hands, legs, and thighs. After consulting with physicians who assured him that the boy was doomed, Pasteur reluctantly decided to administer the vaccine. On July 7, sixty hours after the accident, Joseph Meister was injected with rabbit spinal cord attenuated by fourteen days' drying. In twelve successive inoculations he received stronger and stronger virus until, on July 16, he received an inoculation of still fully virulent spinal cord which had been removed the day before from the body of a rabbit that had died following inoculation with the virus. Joseph Meister exhibited no symptom and returned to Alsace in good health. He later became gatekeeper of the Pasteur Institute. In 1940, fifty-five years after the accident that gave him a lasting place in medical history, he committed suicide rather than open Pasteur's burial crypt for the German invaders [during World War II].

Another Successful Experiment

The second case treated by Pasteur was that of a shepherd, Jean Baptiste Jupille, aged fifteen. Seeing a dog about to attack some children, Jupille had seized his whip in an attempt to drive it away, but was severely bitten; he finally managed to wind his whip around the muzzle of the animal and to crush its skull with his wooden shoe. The dog was subsequently declared rabid, and Jupille was brought to Paris for treatment six days after being bitten. He survived, and his deed was commemorated in a statue which stands today in front of the Pasteur Institute in Paris.

These two dramatic successes encouraged numerous patients to go to Pasteur for treatment after being bitten by animals known or presumed to be rabid. By October 1886, fifteen months after Joseph Meister had first been treated, no fewer than 2,490 persons had received the vaccine. Thus, . . . Pasteur saw his method become an established practice within a short time of its inception, but . . . the rabies treatment was immediately attacked as valueless, and capable of causing the very disease it was designed to control.

Immunization Is Born

I cannot discuss here the very complex technical problems posed by the antirabies treatment in humans. Granted the real difficulties associated with Pasteur's vaccination techniques, it is on much broader issues that his achievements must be judged. Pasteur had demonstrated the possibility of investigating by rigorous techniques the infectious diseases caused by invisible, noncultivable viruses; he had shown that their pathogenic potentialities could be modified by various laboratory artifices; he had established beyond doubt that a solid immunity could be brought about without endangering the life or health of the vaccinated animals. Thanks to the rabies epic, a new science developed which eventually led to the development of vaccination techniques against yellow fever and several other virus diseases. Even more important, immunization became recognized as a general law of nature. All these achievements were the fruit of Pasteur's boldness as an experimenter, and of his mental courage in the face of natural odds and of human opposition.

Louis Pasteur's Rabies Vaccine Is Successful:
July 6, 1885

The Pasteur Institute Opens in Paris

by the *Times*

Louis Pasteur's successful antirabies treatment of July 6, 1885, had several positive effects. In addition to saving human lives, the rabies vaccine offered hope for further developments in antiviral vaccines. This hope served as the inspiration for the Pasteur Institute, which opened in Paris in 1888. The purpose of the Pasteur Institute was to encourage research on rabies vaccination, to develop the study of infectious diseases, and to share Pasteur's findings.

The following November 16, 1888, excerpt from the *Times* of London describes the origin and opening of the Pasteur Institute. The article provides a synopsis of Pasteur's ideas, achievements, and discoveries in the field of biology. The writer of the article believes that the Pasteur Institute will play a significant role in the welfare of humankind. Over a century later, the Pasteur Institute remains an important center for biological research.

The opening of the Pasteur Institute at Paris, on Wednesday, is an event which may well prove to be of lasting importance to the welfare of mankind.

The Institute has owed its origin to the desire of M. Pasteur's admirers to provide him with opportunities for continuing, not only the practice of inoculation for the cure of rabies and of hy-

---

The Times, "The Opening of the Pasteur Institute at Paris," November 16, 1888.

drophobia [another term for rabies, meaning "fear of water"; one symptom of rabies in humans is an aversion to swallowing liquids], but also the researches into the nature of animal poisons by which he has already so greatly enlarged our knowledge of the causes of numerous forms of disease. It is mainly by these researches that many facts which are now familiar have been ascertained, and that the influence of the minute organisms which are commonly called "microbes" has been traced through successive generations obtained by artificial culture, and through a long series of victims to their fatal effects. M. Pasteur commenced his labours in this direction by the discovery that the disease of silkworms, which was so fatally prevalent a few years ago, was due to the introduction of microbes into the affected insect, and to their multiplication within its tissues; and the knowledge thus obtained was soon afterwards applied to an investiga-

Louis Pasteur's rabies vaccine saved lives and offered hope for further developments in antiviral vaccines.

tion into the causes of chicken cholera and of splenic fever in cattle. The last-mentioned disease, which often prevailed in Continental countries, was of the most destructive character, and was so contagious as to attack even healthy animals pastured in a field in which it had long previously existed. M. Pasteur succeeded in identifying and isolating the microbe by which splenic fever was produced; and he showed that this microbe, multiplying in the blood and tissues of the stricken beast, was constantly buried with its carcass, and was liable after long periods to be brought again to the surface by the action of earthworms, to be deposited upon the grass, and once more to exert its deadly influence. M. Pasteur showed, moreover, that the virulence of microbes was greatly dependent upon the conditions under which the multiplication and growth of the organisms occurred; and that these conditions might be so modified by art as to produce either an increased or a diminished intensity of poison. In the latter case, it was found practicable to arrive at a condition which rendered the microbe harmless to life, but which nevertheless left it with the ability to produce some change in the animal body into which it was introduced, which rendered that body insusceptible of the otherwise destructive potency of the more virulent forms. The facts were, indeed, a repetition of the history of vaccination as a preventive of smallpox; but with the difference that the latter discovery was made by the fortuitous application of the experience of milkers, while the former was, from first to last, a result of elaborate experiment, guided by profound sagacity and careful reasoning. Some years have now elapsed since M. Pasteur gave a convincing demonstration of the soundness of his views. Some cattle dead of splenic fever had been buried in a particular field, in which time had been allowed for earthworms to bring the microbes to the surface. A flock of sheep, a portion of which had been protectively inoculated, were turned into the field, and the members of a French Agricultural Society met M. Pasteur to examine them after the lapse of twenty-four hours. As he had predicted, the inoculated sheep were in perfect health, while all those which had not been inoculated were either dead or dying.

Although the possibility of preventing splenic fever was thus completely established, the discovery was one which was chiefly interesting to farmers and breeders, and it failed to receive among the general public as much attention as it deserved. But, nearly four years ago, when M. Pasteur announced that he had discov-

ered the cause of hydrophobia, a malady which had previously
entailed upon all its human victims a horrible death, the announcement at once took the world by storm, and every one who
was bitten hastened to Paris to receive the benefit of the discovery. Even now, in all probability, there is much to learn with regard to the conduct of the inoculations; but M. Pasteur's chief assistant, Dr. Grancher, was able on Wednesday to declare that,
while, previously to 1885, the mortality among persons bitten by
rabid animals was 16 per cent., this had been reduced, among
those inoculated, to 1.34 per cent.; and that, taking the first six
months of the present year, it had been still further reduced to
0.77 per cent., even including the cases of persons in whom
symptoms of disease were apparent before the treatment was
commenced. The material employed is obtained from the spinal
cords of rabid rabbits, and the chief question to be determined is
to what point the reduction of its strength should be carried in order to obtain the greatest security of benefit with the smallest
amount of risk. It cannot be said that risk was at one time wholly
absent; for, in the case of one man who was sent over to Paris
from this country [England] there was reason to believe that the
hydrophobia from which he died was rather a result of the inoculations than of the original bite. In other well-known instances
the treatment has proved ineffectual; but in all of these, as far as
we have been informed, the want of success has been dependent
either upon too much time having been lost before the inoculations were practised, or upon the neglect of precautions which a
more extended experience has shown to be necessary. Notwithstanding occasional failures, the instinct of self-preservation has
attracted no fewer than 5,384 bitten persons to Paris since the
middle of 1885; and, although many objections have been raised
to the method, generally analogous to those which have been
raised against vaccination, the common sense of mankind has
brushed them aside, and has recognized in M. Pasteur a philosopher whose work has been directed with a single aim to the ex-
tinction of some of the greatest scourges to which men and animals have been exposed. One effect of his labours has been to
bring into prominence the enormous total mortality which hydrophobia has produced, especially when we include in the returns statistics obtained from Russia, where the peasantry are
much exposed to the bites of rabid wolves. Instead of being a rare
occurrence, as was once believed, it is now established that death

from hydrophobia, even in this country, occupies a very definite place in the returns of the Registrar-General.

Encouraging Research at the Institute

The proceedings of Wednesday, which were conducted by the President of the Republic, afford gratifying testimony that the political disquietudes of France have in no way diminished her readiness to do honour to her most illustrious *savant*. The funds for the Institute have, indeed, been raised in all countries, the Czar and the Sultan having been contributories; and the total of the subscriptions has reached an amount of more than £100,000, of which about £60,000 have been expanded in erecting and fitting up the building, leaving £40,000 as capital to form the basis of a future income. The French Government, with an enlightened liberality which our own would do well to imitate, has hitherto contributed £5,000 a year towards the expenses of the work in which M. Pasteur has been engaged; but he is very desirous to increase the capital to such an extent as to render the Institute independent of State assistance, and secure of its means of usefulness notwithstanding any possible fluctuations of opinion. The building is arranged with the twofold object of continuing the protective inoculation of persons who have been bitten, and of affording full scope for the prosecution of researches in other directions. It will fulfil, in short, what may be described as the idea of our own "Brown Institution," at which likewise the diseases of animals are studied by competent observers, and in which many researches of great value have been conducted. The chief difference between them is that the English institution is starved for want of funds, its original small endowment having never received any substantial increase; while the French institution may not only be said to start fully equipped upon its career of usefulness, but is also the parent of twenty others in various parts of the world. Besides the President, some of the most eminent men in France assembled to do honour to the great philosopher, and to justify the claim, which they would no doubt be ready to make on behalf of their country, that she has always been in the van of civilization and enlightenment. We rejoice in the well-deserved distinction to which M. Pasteur has attained; but we cannot refrain from expressing some regret that the encouragement of scientific research should be one of the things which they "do better in France" than among ourselves.

The Constitution Announces Japan's Modernity to the World

by James L. McClain

In 1868 Japan underwent a political transformation known as the Meiji Restoration. In January of that year a group of samurai overthrew the reigning shogunate and "restored" power to the emperor Meiji. Following the coup, the samurai leaders abolished the feudal system and established a centralized government. The emperor was an important figure in Japanese history because, according to legend, Japan's emperors were descendants of the Sun. The period of the emperor's reign (1868–1912) is known as the Meiji era.

The Meiji era was characterized by a spirit of reform. The leaders of the coup enacted a series of reforms—political, social, economic, administrative, military, educational, and legal—that transformed Japanese society. These men were motivated by a desire for Japan to play a leading role in world affairs.

One of the most important manifestations of Japanese reform is the constitution that was put into effect on February 11, 1889. As James L. McClain illustrates, the constitution embodied both traditional values and new political ideas. This document developed over the course of seven years and was the subject of much debate among the constitution's architects. McClain writes that Japan

James L. McClain, *Japan: A Modern History*. New York: W.W. Norton & Company, 2002.
Copyright © 2002 by W.W. Norton & Company. Reproduced by permission.

sought political modernity, such as legislature and civic responsibility. Japan looked to other nations as models but also wished to retain its own values of the past, such as obedience to the emperor. McClain describes how statesman Ito Hirobumi and others weighed these considerations and then produced a constitution for the "new" Japan. The document was made public when the emperor "gave" it to the Japanese people in a symbol-laden ceremony.

James L. McClain is a professor of history at Brown University. He has written three books about Japanese history, including *Japan: A Modern History,* from which this excerpt is taken.

For Itō Hirobumi [a prominent Japanese statesman who served as premier four times] it was the defining, crowning moment in what his colleagues hailed as a lifetime of achievement. On the morning of February 11, 1889, Itō approached the emperor of Japan, whose throne rested atop a red-carpeted dais in the refurbished reception chamber of the imperial compound. Resplendent in formal Western-style dress, and in full view of a gallery filled to overflowing with Japan's peerage, ministers of state, prefectural governors, high-ranking military officials, and foreign diplomats, Itō offered up to the emperor a scroll inscribed with Japan's new constitution. Without either reading or examining the document, the emperor promptly handed it down to the prime minister, Kuroda Kiyotaka, who "received it with a deep reverence." Turning, the emperor "nodded" and departed the hall to the strains of "Kimigayo," Japan's newly composed, although still unofficial, national anthem. Outside, bells rang and cannons roared. In a ceremony that had lasted just a few brief moments, Emperor Meiji had bestowed upon his subjects the Constitution of the Empire of Japan.

The Coming Together of Ideas

As the symbolism of the ceremony suggested, however, the monarch had little to do with the actual preparation of the new "fundamental law of the state"; its chief architect was Itō, who had spent the better part of a decade pondering and crafting its every phrase. Even so, Itō had not worked in calm isolation. During the years he and other Meiji leaders spent contemplating and drafting a constitution, a storm of controversy swirled around them as many of their countrymen sought to influence the shape

of the new law of the land, whose promulgation, nearly everyone agreed, was essential if the Japanese were to travel the road
of progress, as promised by the slogan "Civilization and Enlightenment," and if Japan were to prove to the West that it deserved respect and equal treatment within the community of
modern nations. . . .

As the oligarchs moved toward constitutionalism, they sought
to define a political orthodoxy and promote ideals of civic virtue
that would convert the residents of the nearly three hundred scattered domains that had existed before 1871 into *kokumin*, "national citizens" unquestioningly loyal to the new state and
bonded together around a core of common political values. Itō
and his colleagues had reason for concern: The optimism of the
bun-mei kaika [civilization and enlightenment] epoch had fa-
vorably disposed many Japanese toward the regime, but the oligarchs scarcely could hope to complete their slate of reforms if
they were unable to establish the legitimacy of the new constitutional state. Consequently, all national leaders in the 1870s and
1880s lent a hand to the conscious enterprise of cobbling together
a package of ideological values and civic ideals that would enable the ordinary Japanese to make coherent sense of the evolv-
ing political arrangements engineered by the young regime and
to construe for themselves a meaningful place within the new
Japan of the Meiji era.

Each oligarch seemed to have his own individual contribution
to make to the construction of a suitable political ethos. Itō spoke
wistfully of a "common folk" who were peaceful and docile,
honest, industrious, happy to sacrifice their individual interests
for the sake of their neighbors—"splendid material with which
to build up a strong nation," as he put it. [Education minister] Inoue Kowashi, for his part, preferred to stress loyalty and patriotism as the key ingredients of citizenship; without those virtues
there could be no national unity, and little that was worthwhile
would be accomplished. [Soldier and statesman] Yamagata Aritomo certainly shared Inoue's convictions, and he was also a bitter foe of private interest, represented in the political realm by
the opposition's political associations, since they threatened to
drown a collective, harmonious, and nonpartisan struggle for
progress in a sea of divisiveness, disunity, and chaos.

As the oligarchs worked through their ideas, they concluded
that the emperor should provide the nucleus for a political or-

thodoxy, just as he would constitute the heart of the new political system. But that message needed to be heralded with colorful banners and loud trumpets since as [political leader Tomomi] Iwakura [who played a key role in the Meiji Restoration of 1868] noted in the 1870s, the shogun had overshadowed the emperor during the Tokugawa era, with the consequence that "the imperial will is as yet unknown in the remote and distant parts" of the realm. Thus, the young Meiji leadership sent the emperor off to visit the hinterlands of Japan so that "all the people of the nation will turn their eyes and see the greatness of the emperor's conduct." In the two and a half centuries of Tokugawa rule Heavenly Sovereigns ventured outside their home city of Kyoto just three times; in the forty-five years of his enthronement, the Meiji emperor made 102 excursions, including the six Great Circuits of 1872, 1876, 1878, 1880, 1881, and 1885. Traveling by every conveyance available, the handsome young monarch covered thousands of miles, visited every major island, accepted the well-wishes of the farmers who came in from their fields to offer a humble bow, and lodged with local notables, usually entering their residences through a newly erected gate and taking his comfort upon an imperial privy especially installed for the occasion. Attending the emperor, not surprisingly, were men like [political leaders] Kido, Itō, Ōkubo, Kuroda, and Iwakura, all of whom spent some time on the road. The processionals, they agreed, were a grand scheme; the imperial presence informed "the people of the emperor's great virtue" and represented "the opportunity of displaying direct imperial rule in the flesh, thus dispelling misgivings" about proposals for a constitutional monarchy.

Tradition and Orthodoxy

As the emperor traversed his islands, many outside government began to voice opinions about political orthodoxy that echoed the values preached by the oligarchs. Fukuchi Gen'ichirō was among the journalists who lined up behind Itō and the other oligarchs. Although scandalized by what he called "the government's slipshod ways in the Hokkaidō land deal," the editor of the *Tōkyō nichinichi shinbun* agreed that constitutionalism was desirable and that the emperor ought to be the sovereign linchpin of the new system. Sovereignty, Fukuchi reiterated in dozens of editorials in the early 1880s, had to be decided in accordance with each nation's historical circumstances. To underscore his point, he wrote about Japan's *kokutai*, adopting the emotion-laden term, usually rendered as "national essence" or "national polity," that had been at the center of anti-Tokugawa rhetoric in the 1850s and 1860s. Japan's *kokutai*, Fukuchi reminded his readers, rested on the principle that a divine line of emperors had reigned, and reigned wisely, since the dawn of Japanese history. As Japan entered the modern world, he continued, it was appropriate that the structures of governance change, but the essence of authority, the *kokutai*, must be maintained unchanged. "The gods," he wrote passionately, "decreed at the time of the first ascension to the throne that the affairs of state" should be entrusted to the emperor. Thus, even within a constitutional framework, it was necessary that "all government be carried out by imperial will." By adhering to the "fixed principle" of the *kokutai*, he argued, the government would ensure itself of the allegiance of the citizenry as the emperor led everyone forward on a steady and peaceful evolution into modernity.

Certain intellectuals also attempted to articulate a political ethos that bound the new world of representative government within a framework of traditional virtues. Hozumi Yatsuka, still a student and only approaching the threshold of a long career as a constitutional scholar, contributed in 1882 a series of essays to Fukuchi's *Tōkyō nichinichi shinbun* that expressed his abhorrence of political parties since they placed the egocentric interests of private citizens above the needs of the state. In many instances, Hozumi wrote, "the liberty of the individual must be sacrificed for the sake of society as a whole," and the best defense against the tyranny of the masses, he asserted, was to place sovereignty with the emperor, who would then rule through a strong, independent bureaucracy that enjoyed secure tenure in office, public respect, and generous salaries.

Even Confucianists [followers of the teachings of sixth-century B.C. Chinese philosopher Confucius], on the defensive throughout much of the era of Civilization and Enlightenment, elbowed their way into the dialogue about political ethos. By the beginning of the 1880s such men as Motoda Nagazane, tutor to the emperor, and Nishimura Shigeki, moralist and author, had become alarmed by the oligarchs' enthusiasm for Western-style reform. In Nishimura's judgment, Itō was not doing nearly enough to promote traditional morality, and his flirtation with Prussian ideas of constitutionalism were bound to fail because the oligarch was merely transplanting things Western into Japan in helter-skelter fashion. The irascible Motoda was characteristically blunt; Itō and his colleagues, he fumed, seemed determined to convert the "Japanese into painted replicas of Europeans and Americans." The Confucianists' prescription was simple: Return to time-tested values. Morality, Nishimura wrote, was the foundation of everything from one's personal life to the structure of government, and to be moral, he stressed, was to observe "the spirit of loyalty, filial piety, honor, duty, valor and shame which had been the traditional foundation of our country since ancient times." The point was obvious, Motoda concurred. Any new political arrangement had to orbit tightly around the emperor since he was the model of virtue, the custodian of the ancestral precepts transmitted across the ages.

Merging Past and Present Values

The remarkable synergism arising from the multiple concerns about establishing an appropriate political ethos was not lost on Itō as he pondered his constitution. Indeed, by the time he drew together his drafting committee in the middle of the 1880s, he had developed views about the emperor that diverged significantly from those being urged upon him by the European experts. [German legal consultants] Mosse and Roesler subscribed to the notion of social monarchy, in which an abstract state, embodied in the person of the monarch, stood above society and carried out reforms for the benefit of the citizenry. However, they went on, Europe's historical past bore the scars of those occasions when unrestrained royal prerogatives had distilled into personal absolutism. Thus law and legal institutions were necessary in the present day so that the throne would exercise its power in a manner that fostered the well-being of all social classes.

Itō agreed that reliance on law and restraints against the arbitrary exercise of power by strong monarchs were essential components of any modern constitutional system. He also concurred with his chief lieutenant, Inoue Kowashi, that while the emperor "governs the people, he does not personally administer the government," a function more appropriately entrusted to ministers of state. Yet, as Itō later also reminisced, "the Crown was an institution far more deeply rooted in the national sentiment and in our history than in other countries. It was indeed the very essence of a once theocratic State." As a consequence, he declared, it was

evident "that the mere imitation of foreign models would not suf-
fice, for there were historical peculiarities of our country which
had to be taken into consideration." Specifically, he concluded,
it was imperative that the constitution "take care to safeguard the
future realness or vitality of these imperial prerogatives, and not
let the institution degenerate into an ornamental crowning piece
of the edifice." In the end Itō and Inoue framed a constitution that
recalled a cardinal principle of the Japanese past: The emperor
reigned but seldom ruled. The final document placed considerable powers in the hands of state ministers, but Itō and Inoue personally wrote the articles that dealt with the imperial house,
adopting language to make it abundantly clear, as Itō put it, that
"the first principle of our constitution is the respect for the sovereign rights of the emperor." Foreign models were useful, perhaps even essential, but Itō was determined that Japan's quest for
political modernity proceed on its own terms and in a manner
congruent with the most hallowed traditions of its past.

To convey better their sense of civil ethics to the population at
large, the oligarchs promulgated two well-publicized rescripts.
The first, the Imperial Rescript to Soldiers and Sailors, flowed
chiefly from the pens of Inoue Kowashi and Yamagata Aritomo,
with some last-second stylistic flourishes added by the progov-
ernment journalist Fukuchi Gen'ichirō. On January 4, 1882, the
emperor presented the rescript personally to the army minister at
the palace, a ceremony intended to symbolize the throne's direct
authority over the military. In its very first article, the document
proclaimed "loyalty" to be the "essential duty" of the soldier and
sailor and asked rhetorically, "Who that is born in this land can
be wanting in the spirit of grateful service to it?" The ideal of service to the throne, the rescript then admonished, could be cor-
rupted by the selfishness and divisiveness of party politics. Thus,
to be loyal meant to stand outside the fray of the political arena:
"Remember that, as the protection of the state and the maintenance of its power depend upon the strength of its arms, the
growth or decline of this strength must affect the nation's destiny
for good or evil; neither be led astray, therefore, by popular opinions nor meddle in politics, but with single heart fulfill your essential duty of loyalty, and bear in mind that duty is weightier
than a mountain, while death is lighter than a feather."

The second rescript sought a wider audience, the entire youth
of the nation. In the autumn of 1890, just before the convocation

of the first Imperial Diet, Prime Minister Yamagata instructed Inoue Kowashi and Motoda Nagazane to compose a document for
dissemination to all schools articulating the bedrock ethos that
supported the new state, with its brand-new constitution and system of participatory government. Encoding a political ideology
within a statement outlining guiding principles for education, the
Imperial Rescript on Education opened with the declaration that
Japan's unique *kokutai*, "the fundamental character of Our Empire" in the official English translation, sprang from the intimate
bonds that from time immemorial had united Japan's benevolent
rulers with their loyal subjects: "Our Imperial Ancestors have
founded Our Empire on a basis broad and everlasting, and have
deeply and firmly planted virtue; Our subjects ever united in loyalty and filial piety have from generation to generation illustrated
the beauty thereof." From that construction flowed a variety of
prescriptions about the present: The Japanese were to be filial to
parents, affectionate to siblings, true to friends. Above all else,
the rescript exhorted students to be "good and faithful subjects,"
and it instructed them to "advance public good and promote
common interests; always respect the Constitution and observe
the laws; [and] should emergency arise, offer yourselves coura-
geously to the State; and thus guard and maintain the prosperity
of Our Imperial Throne coeval with heaven and earth." Throughout the country, school principals required their charges to mem-
orize the text, and its solemn recitation, together with obeisance
to a photograph of the emperor and the singing of "Kimigayo,"
became an obligatory part of all important school ceremonies.

The symbolism of the imperial presence, "coeval with heaven
and earth," proved enormously compelling, and the power of its
message appealed to even the most ardent sympathizers with the
Popular Rights Movement. Of course critics blasted the Satchō
oligarchs as self-seeking, opportunistic, and corrupt, the very an-
tithesis of what they preached, but during the 1880s nearly all the
regime's foes came to accept the basic premises of imperial constitutionalism. The polemics held a high pitch, but none of the
constitutional drafts proposed by the government's opponents
during the latter half of the decade failed to place the emperor at
the center of the political world. Similarly, people across a broad
range of the political spectrum embraced the oligarchs' notions
of acceptable political discourse and civic morality: obedience
to emperor; loyalty to nation; the sanctity of rule by a disinter-

ested, well-intentioned bureaucracy presided over by imperially appointed ministers; selfless service; denial of divisiveness inherent in interest group politics; and sublimation of private interest to the greater collective good of the nation. Such rhetoric did not yet constitute an official "state ideology"—that would be formulated more clearly in the 1890s and the opening decades of the twentieth century—but the notions of imperial orthodoxy, civic responsibility, and good citizenship were extremely useful to the oligarchs, for they served as political and moral bonds that defined a national citizenry, drawing the *kokumin* together and readying them to accept Mr Itō's new constitution.

The Meaning of the Constitution

For Itō and his colleagues, the Constitution of the Empire of Japan announced Japan's modernity to the world, exemplified the nation's enlightened progress, and established new concepts of political authority, all delicately wrapped in the legacy of the past and legitimated by ancient values. Symbolism was important to Itō Hirobumi, and the promulgation of the constitution demonstrated the full measure of his creative ingenuity. The day chosen for its unveiling was February 11, a national holiday commemorating the mythological founding of the country in 660 B.C.E. by Jimmu, the legendary first emperor. Early that morning in 1889 the Meiji emperor, cloaked in ancient court garments, performed Shinto rituals before a special shrine erected deep inside the imperial compound. There he informed his ancestors of the new "fundamental law of the state." He was granting the constitution, he told them, out of "consideration of the progressive tendency of the course of human affairs and in parallel with the advance of civilization." His goal, he explained, was "to give greater firmness to the stability of Our country and to promote the welfare of all the people within the boundaries of Our dominions," and his actions, he assured them, would guarantee "that the observance of Our laws shall continue to the remotest ages of time." That ceremony completed, the emperor changed into Western clothing and appeared in the European-style reception hall. There he received the constitution from Itō and handed it down to Prime Minister Kuroda, a sign to all that the constitution was being bestowed upon the nation by a gracious sovereign.

The document itself was clearly the repository of tradition and the embodiment of modernity. The first seventeen articles, grouped together under the heading "Chapter 1," concerned the emperor. The very first article established that "The Empire of Japan shall be reigned over and governed by a line of Emperors unbroken for ages eternal," while the following clauses declared him "sacred and inviolable" with full and sole possession of "the rights of sovereignty." The remainder of Chapter 1 specified the monarch's duties. Of particular importance, it empowered the emperor to open, close, prorogue, and dissolve the Imperial Diet. In addition, according to Article 6, the emperor "gives sanction to laws, and orders them to promulgated and executed." Article 8 extended that prerogative, authorizing the throne to issue "imperial Ordinances in the place of law" when the Diet was not sitting, and other provisions designated the emperor the supreme commander of the armed forces and entitled him to declare war, make peace and conclude treaties. However, the constitution also guarded against the arbitrary exercise of power by circumscribing imperial prerogatives; one clause, for instance, specified that "Imperial Ordinances are to be laid before the Imperial Diet at its next session" and further declared that no ordinance could "in any way alter any of the existing laws."

Chapter 2 took up "the rights and duties of subjects." The two chief duties were to pay taxes and to serve in the army and navy when called upon. The rights were more numerous: to have "the liberty of abode" and be free from unlawful entry or search; to be entitled to a trial by judge; and to enjoy the "inviolate" right of property, "freedom of religion," and "the liberty of speech, writing, publication, public meetings and associations." Most of those rights and privileges were not granted unconditionally but, rather were qualified by such phrases as "within the limits of the law" and when "not antagonistic to their duties as subjects." Moreover, any right could be abrogated in times of national emergency. It was essential, Itō noted in his commentaries, "to make exceptional provisions to meet exceptional contingencies. For it must be remembered that the ultimate aim of a State is to maintain its existence." As a consequence, according to Itō, "in times of danger, the State will have to sacrifice without hesitation part of the law and of the rights of the subjects."

The next several chapters established a bicameral legislature and outlined the duties of ministers of state and the Privy Council. Both ministers and councillors were to provide advice to the throne, and in another check on imperial power, no law, ordinance, or rescript could take effect unless countersigned by the appropriate minister. The newly enfranchised Imperial Diet consisted of a House of Peers, composed of nobles and imperial family members appointed by the emperor, and a House of Representatives, elected by the people. Either house could initiate bills and make representations to the cabinet, and all legislation required the consent of both bodies. The constitution also empowered the Diet to vote on an annual state budget. But several articles qualified that important right. Article 67 stipulated that "already fixed expenditures based . . . upon the powers appertaining to the Emperor . . . shall be neither rejected nor reduced by the Imperial Diet." That clause reflected the oligarchs' fear that a cantankerous, popularly elected Diet might someday jeopardize national security by trimming appropriations for the armed forces, which fell into the category of "fixed expenses . . . appertaining to the Emperor." Article 71 circumscribed the Diet's budgetary powers even further. It stipulated that whenever the legislature did not pass a budget, "the Government shall carry out the Budget of the preceding year."

A Modern Japan

Itō did not escape unbloodied from the fierce debates over constitutionalism that raged during the early Meiji era, but by mixing concessions with legal suppression and ideological suasion, he and his fellow oligarchs ultimately achieved their goals. The Constitution of the Empire of Japan reflected the ideas set forth in the "general principles" that Iwakura articulated at the beginning of the 1880s: It established mechanisms for sharing power with elected representatives of the nation's citizenry, but at the same time it drew boundaries around the exercise of legislative initiative, reserved sovereignty for the emperor, and delegated preponderant political authority to state ministers appointed by the throne.

The promulgation of the constitution also marked an important step toward fulfilling another ambition: acceptance into the comity of the world's advanced nations. Almost from the day they seized power, the Meiji oligarchs had been determined to build internal strength and foster national unity so that they could go forward with the task of creating a modern, powerful country, worthy of respect in the eyes of the West. Only through such a process, they believed, could Japan maintain its national independence, revise the humiliating unequal treaties, and, as Yamagata once put it, "preserve the nation's rights and advantages among the powers." The constitution helped Japan earn the international respect that it desperately sought and to attain great power status. In his later years Itō expressed his satisfaction with the constitutional experiment. "Looking backwards over sixteen years of its operation," he wrote, "it is not without a certain degree of gratification that I feel the experiment has been, in spite of its many shortcomings, upon the whole a success."

One might well forgive Itō the conceits of his accomplishments, but in fact, as the present unfolded into the future, politics under the Meiji constitution evolved in ways that he could scarcely have imagined. For all its deference to imperial prerogatives, the modernistic fundamental law of the land empowered new elites, including the prime minister and his cabinet, the Privy Council, the Diet, and the military and civilian bureaucracies, all of which would compete vigorously for a share of political influence in the years and decades ahead. The ebb and flow of that contest among multiple elites ultimately took Japan in directions totally unforeseen by Itō when he approached the red-carpeted dais that February morning in 1889 to present his constitution to the emperor of Japan.

**A Constitution Modernizes Japan's
Government: February 11, 1889**

The Emperor and His People

by Hirobumi Ito

Japan's constitution was presented to the people on February 11, 1889. The date chosen for this special event was significant. It was a national holiday honoring Japan's legendary founding by an emperor in 660 B.C. By revealing the constitution on this day, Japan's leaders announced a "new" Japan.

The document, which was drafted by statesman and eventual prime minister Hirobumi Ito, among others, guided Japan's politics from 1889 to 1945. The most prominent figure in Japan, according to both tradition and the new constitution, is the emperor. The constitution begins with an imperial oath, an imperial rescript, and a preamble, all of which invoke Japan's glorious ancestors.

W e, the Successor to the prosperous Throne of Our Predecessors, do humbly and solemnly swear to the Imperial Founder of Our House and to Our other Imperial Ancestors that, in pursuance of a great policy co-extensive with the Heavens and with the Earth, We shall maintain and secure from decline the ancient form of government.

In consideration of the progressive tendency of the course of human affairs and in parallel with the advance of civilization, We deem it expedient, in order to give clearness and distinctness to the instructions bequeathed by the Imperial Founder of Our House and by Our other Imperial Ancestors, to establish funda-

Hirobumi Ito, *Commentaries on the Constitution of the Empire of Japan*, translated by Miyoji Ito. Tokyo: Igirisu-horitsu gakko, 1889.

mental laws formulated into express provisions of law, so that, on the one hand, Our Imperial posterity may possess an express guide for the course they are to follow, and that, on the other, Our subjects shall thereby be enabled to enjoy a wider range of action in giving Us their support, and that the observance of Our laws shall continue to the remotest ages of time. We will thereby to give greater firmness to the stability of Our country and to promote the welfare of all the people within the boundaries of Our dominions; and We now establish the Imperial House Law and the Constitution. These Laws come to only an exposition of grand precepts for the conduct of the government, bequeathed by the Imperial Founder of Our House and by Our other Imperial Ancestors. That we have been so fortunate in Our reign, in keeping with the tendency of the times, as to accomplish this work, We owe to the glorious Spirits of the Imperial Founder of Our House and of Our other Imperial Ancestors.

We now reverently make Our prayer to Them and to Our Illustrious Father, and implore the help of Their Sacred Spirits, and make to Them solemn oath never at this time nor in the future to fail to be an example to our subjects in the observance of the Laws hereby established.

May the heavenly Spirits witness this Our solemn Oath.

Imperial Rescript on the Promulgation of the Constitution

Whereas We make it the joy and glory of Our heart to behold the prosperity of Our country, and the welfare of Our subjects, We do hereby, in virtue of the Supreme power We inherit from Our Imperial Ancestors, promulgate the present immutable fundamental law, for the sake of Our present subjects and their descendants.

The Imperial Founder of Our House and Our other Imperial ancestors, by the help and support of the forefathers of Our subjects, laid the foundation of Our Empire upon a basis, which is to last forever. That this brilliant achievement embellishes the annals of Our country, is due to the glorious virtues of Our Sacred Imperial ancestors, and to the loyalty and bravery of Our subjects, their love of their country and their public spirit. Considering that Our subjects are the descendants of the loyal and good subjects of Our Imperial Ancestors, We doubt not but that Our subjects will be guided by Our views, and will sympathize with all Our endeavors, and that, harmoniously cooperating together, they will share with Us Our hope of making manifest the glory of Our country, both at home and abroad, and of securing forever the stability of the work bequeathed to Us by Our Imperial Ancestors.

Preamble [or Edict] (Joyu)

Having, by virtue of the glories of Our Ancestors, ascended the throne of a lineal succession unbroken for ages eternal; desiring to promote the welfare of, and to give development to the moral and intellectual faculties of Our beloved subjects, the very same that have been favored with the benevolent care and affectionate vigilance of Our Ancestors; and hoping to maintain the prosperity of the State, in concert with Our people and with their support, We hereby promulgate, in pursuance of Our Imperial Rescript of the 12th day of the 10th month of the 14th year of Meiji, a fundamental law of the State, to exhibit the principles, by which We are guided in Our conduct, and to point out to what Our descendants and Our subjects and their descendants are forever to conform.

The right of sovereignty of the State, We have inherited from Our Ancestors, and We shall bequeath them to Our descendants. Neither We nor they shall in the future fail to wield them, in accordance with the provisions of the Constitution hereby granted.

We now declare to respect and protect the security of the rights and of the property of Our people, and to secure to them the complete enjoyment of the same, within the extent of the provisions of the present Constitution and of the law.

The Imperial Diet shall first be convoked for the 23rd year of Meiji and the time of its opening shall be the date, when the present Constitution comes into force.

When in the future it may become necessary to amend any of the provisions of the present Constitution, We or Our successors shall assume the initiative right, and submit a project for the same to the Imperial Diet. The Imperial Diet shall pass its vote upon it, according to the conditions imposed by the present Constitution, and in no otherwise shall Our descendants or Our subjects be permitted to attempt any alteration thereof.

Our Ministers of State, on Our behalf, shall be held responsible for the carrying out of the present Constitution, and Our present and future subjects shall forever assume the duty of allegiance to the present Constitution.

Chapter 1: The Emperor

Article 1. The Empire of Japan shall be reigned over and governed by a line of Emperors unbroken for ages eternal.

Article 2. The Imperial Throne shall be succeeded to by Imperial male descendants, according to the provisions of the Imperial House Law.

Article 3. The Emperor is sacred and inviolable.

Article 4. The Emperor is the head of the Empire, combining in Himself the rights of sovereignty, and exercises them, according to the provisions of the present Constitution.

Article 5. The Emperor exercises the legislative power with the consent of the Imperial Diet.

Article 6. The Emperor gives sanction to laws, and orders them to be promulgated and executed.

Article 7. The Emperor convokes the Imperial Diet, opens, closes, and prorogues it, and dissolves the House of Representatives.

Article 8. The Emperor, in consequence of an urgent necessity to maintain public safety or to avert public calamities, issues, when the Imperial Diet is not sitting, Imperial ordinances in the place of law.

(2) Such Imperial Ordinances are to be laid before the Imperial Diet at its next session, and when the Diet does not approve the said Ordinances, the Government shall declare them to be invalid for the future.

Article 9. The Emperor issues or causes to be issued, the Ordinances necessary for the carrying out of the laws, or for the maintenance of the public peace and order, and for the promotion of the welfare of the subjects. But no Ordinance shall in any way alter any of the existing laws.

Article 10. The Emperor determines the organization of the different branches of the administration, and salaries of all civil and military officers, and appoints and dismisses the same. Exceptions especially provided for in the present Constitution or in other laws, shall be in accordance with the respective provisions (bearing thereon).

Article 11. The Emperor has the supreme command of the Army and Navy.

Article 12. The Emperor determines the organization and peace standing of the Army and Navy.

Article 13. The Emperor declares war, makes peace, and concludes treaties.

Article 14. The Emperor declares a state of siege.

(2) The conditions and effects of a state of siege shall be determined by law.

Article 15. The Emperor confers titles of nobility, rank, orders and other marks of honor.

Article 16. The Emperor orders amnesty, pardon, commutation of punishments and rehabilitation.

Article 17. A Regency shall be instituted in conformity with the provisions of the Imperial House Law.

(2) The Regent shall exercise the powers appertaining to the Emperor in His name.

Chapter 2: Rights and Duties of Subjects

Article 18. The conditions necessary for being a Japanese subject shall be determined by law.

Article 19. Japanese subjects may, according to qualifications determined in laws or ordinances, be appointed to civil or military or any other public offices equally.

Article 20. Japanese subjects are amenable to service in the Army or Navy, according to the provisions of law.

Article 21. Japanese subjects are amenable to the duty of paying taxes, according to the provisions of law.

Article 22. Japanese subjects shall have the liberty of abode and of changing the same within the limits of the law.

Article 23. No Japanese subject shall be arrested, detained, tried or punished, unless according to law.

Article 24. No Japanese subject shall be deprived of his right of being tried by the judges determined by law.

Article 25. Except in the cases provided for in the law, the house of no Japanese subject shall be entered or searched without his consent.

Article 26. Except in the cases mentioned in the law, the secrecy of the letters of every Japanese subject shall remain inviolate.

Article 27. The right of property of every Japanese subject shall remain inviolate.

(2) Measures [deemed] necessary to be taken for the public benefit shall be any [measures that are] provided for by law.

Article 28. Japanese subjects shall, within limits not prejudicial to peace and order, and not antagonistic to their duties as subjects, enjoy freedom of religious belief.

Article 29. Japanese subjects shall, within the limits of law, enjoy the liberty of speech, writing, publication, public meetings and associations.

Article 30. Japanese subjects may present petitions, by observing the proper forms of respect, and by complying with the rules specially provided for the same.

Article 31. The provisions contained in the present Chapter shall not affect the exercises of the powers appertaining to the Emperor, in times of war or in cases of a national emergency.

Article 32. Each and every one of the provisions contained in the preceding Articles of the present Chapter, that are not in conflict with the laws or the rules and discipline of the Army and Navy, shall apply to the officers and men of the Army and of the Navy.

Chapter 3: The Imperial Diet

Article 33. The Imperial Diet shall consist of two Houses, a House of Peers and a House of Representatives.

Article 34. The House of Peers shall, in accordance with the ordinance concerning the House of Peers, be composed of the members of the Imperial Family, of the orders of nobility, and of those who have been nominated thereto by the Emperor.

Article 35. The House of Representatives shall be composed of members elected by the people, according to the provisions of the law of Election.

Article 36. No one can at one and the same time be a Member of both Houses.

Article 37. Every law requires the consent of the Imperial Diet.

Article 38. Both Houses shall vote upon projects of law submitted to it by the Government, and may respectively initiate projects of law.

Article 39. A Bill, which has been rejected by either the one or the other of the two Houses, shall not be brought in again during the same session.

Article 40. Both Houses can make representations to the Government, as to laws or upon any other subject. When, however, such representations are not accepted, they cannot be made a second time during the same session.

Article 41. The Imperial Diet shall be convoked every year.

Article 42. A session of the Imperial Diet shall last during three months. In case of necessity, the duration of a session may be prolonged by the Imperial Order.

Article 43. When urgent necessity arises, an extraordinary session may be convoked in addition to the ordinary one.

(2) The duration of an extraordinary session shall be determined by Imperial Order.

Article 44. The opening, closing, prolongation of session and prorogation of the Imperial Diet, shall be effected simultaneously for both Houses.

(2) In case the House of Representatives has been ordered to dissolve, the House of Peers shall at the same time be prorogued.

Article 45. When the House of Representatives has been ordered to dissolve, Members shall be caused by Imperial Order to be newly elected, and the new House shall be convoked within five months from the day of dissolution.

Article 46. No debate can be opened and no vote can be taken in either House of the Imperial Diet, unless not less than one-third of the whole number of Members thereof is present.

Article 47. Votes shall be taken in both Houses by absolute majority. In the case of a tie vote, the President shall have the casting vote.

Article 48. The deliberations of both Houses shall be held in public. The deliberations may, however, upon demand of the Government or by resolution of the House, be held in secret sitting.

Article 49. Both Houses of the Imperial Diet may respectively present addresses to the Emperor.

Article 50. Both Houses may receive petitions presented by subjects.

Article 51. Both Houses may enact, besides what is provided for in the present Constitution and in the Law of the Houses, rules necessary for the management of their internal affairs.

Article 52. No Member of either House shall be held responsible outside the respective Houses, for any opinion uttered or for any vote given in the House. When, however, a Member himself has given publicity to his opinions by public speech, by documents in print or in writing, or by any other similar means, he shall, in the matter, be amenable to the general law.

Article 53. The Members of both Houses shall, during the session, be free from arrest, unless with the consent of the House, except in cases of flagrant delicts, or of offenses connected with a state of internal commotion or with a foreign trouble.

Article 54. The Ministers of State and the Delegates of the Government may, at any time, take seats and speak in either House.

Chapter 4: The Ministers of State and the Privy Council

Article 55. The respective Ministers of State shall give their advice to the Emperor, and be responsible for it.

(2) All Laws, Imperial Ordinances, and Imperial Rescripts of whatever kind, that relate to the affairs of the state, require the countersignature of a Minister of State.

Article 56. The Privy Councillors shall, in accordance with the provisions for the organization of the Privy Council, deliberate upon important matters of State when they have been consulted by the Emperor.

Chapter 5: The Judicature

Article 57. The Judicature shall be exercised by the Courts of Law according to law, in the name of the Emperor.

(2) The organization of the Courts of Law shall be determined by law.

Article 58. The judges shall be appointed from among those, who possess proper qualifications according to law.

(2) No judge shall be deprived of his position, unless by way of criminal sentence or disciplinary punishment.

(3) Rules for disciplinary punishment shall be determined by law.

Article 59. Trials and judgments of a Court shall be conducted publicly. When, however, there exists any fear, that such publicity may be prejudicial to peace and order, or to the maintenance of public morality, the public trial may be suspended by provisions of law or by the decision of the Court of Law.

Article 60. All matters that fall within the competency of a special Court, shall be specially provided for by law.

Article 61. No suit at law, which relates to rights alleged to have been infringed by the illegal measures of the administrative authorities, and which shall come within the competency of the Court of Administrative Litigation specially established by law, shall be taken cognizance of by Court of Law.

Chapter 6: Finance

Article 62. The imposition of a new tax or the modification of the rates (of an existing one) shall be determined by law.

(2) However, all such administrative fees or other revenue having the nature of compensation shall not fall within the category of the above clause.

(3) The raising of national loans and the contracting of other liabilities to the charge of the National Treasury, except those that are provided in the Budget, shall require the consent of the Imperial Diet.

Article 63. The taxcs levied at present shall, in so far as they are not remodelled by a new law, be collected according to the old system.

Article 64. The expenditure and revenue of the State require the consent of the Imperial Diet by means of an annual Budget.

(2) Any and all expenditures overpassing the appropriations set forth in the Titles and Paragraphs of the Budget, or that are not provided for in the Budget, shall subsequently require the approbation of the Imperial Diet.

Article 65. The Budget shall be first laid before the House of Representatives.

Article 66. The expenditures of the Imperial House shall be defrayed every year out of the National Treasury, according to the present fixed amount for the same, and shall not require the consent thereto of the Imperial Diet, except in case an increase thereof is found necessary.

Article 67. Those already fixed expenditures based by the Constitution upon the powers appertaining to the Emperor, and such expenditures as may have arisen by the effect of law, or that appertain to the legal obligations of the Government, shall be neither rejected nor reduced by the Imperial Diet, without the concurrence of the Government.

Article 68. In order to meet special requirements, the Government may ask the consent of the Imperial Diet to a certain amount as a Continuing Expenditure Fund, for a previously fixed number of years.

Article 69. In order to supply deficiencies, which are unavoidable, in the Budget, and to meet requirements unprovided for in the same, a Reserve Fund shall be provided in the Budget.

Article 70. When the Imperial Diet cannot be convoked, owing to the external or internal condition of the country, in case of urgent need for the maintenance of public safety, the Government may take all necessary financial measures, by means of an Imperial Ordinance.

(2) In the case mentioned in the preceding clause, the matter shall be submitted to the Imperial Diet at its next session, and its approbation shall be obtained thereto.

Article 71. When the Imperial Diet has not voted on the Budget, or when the Budget has not been brought into actual existence, the Government shall carry out the Budget of the preceding year.

Article 72. The final account of the expenditures and revenues of the State shall be verified and confirmed by the Board of Audit, and it shall be submitted by the Government to the Imperial Diet, together with the report of verification of the said board.

(2) The organization and competency of the Board of Audit shall of determined by law separately.

Chapter 7: Supplementary Rules

Article 73. When it has become necessary in future to amend the provisions of the present Constitution, a project to the effect shall be submitted to the Imperial Diet by Imperial Order.

(2) In the above case, neither House can open the debate, unless not less than two-thirds of the whole number of Members are present, and no amendment can be passed, unless a majority of not less than two-thirds of the Members present is obtained.

Article 74. No modification of the Imperial House Law shall be required to be submitted to the deliberation of the Imperial Diet.

(2) No provision of the present Constitution can be modified by the Imperial House Law.

Article 75. No modification can be introduced into the Constitution, or into the Imperial House Law, during the time of a Regency.

Article 76. Existing legal enactments, such as laws, regulations, Ordinances, or by whatever names they may be called, shall, so far as they do not conflict with the present Constitution, continue in force.

(2) All existing contracts or orders, that entail obligations upon the Government, and that are connected with expenditure, shall come within the scope of Article 67.

Hitler's Place in History

by John Lukacs

On April 20, 1889, Adolf Hitler was born in Braunau, Austria. This son of an Austrian customs official from humble beginnings became one of the most significant—and reviled—historical figures of the twentieth century.

Hitler's career as a politician began after Germany's 1918 defeat in World War I. A German war veteran, the anti-Semitic Hitler blamed "the Jews" for the defeat. He joined a radical political party that in 1920 took the name the National Socialist German Workers' (Nazi) Party, and he became the party's chairman in 1921.

In November 1923 Hitler staged the Beer Hall Putsch. This unsuccessful attempt to overthrow the government led to a brief prison sentence, after which he reconstituted the Nazi Party. He then adopted a new tactic: to overthrow the government by constitutional means. Nazis ran for the legislature and won a number of seats in the late 1920s and early 1930s. In 1932 Hitler ran for president of Germany but lost to World War I general Paul von Hindenburg.

Hitler was appointed chancellor on January 30, 1933, but his chancellorship soon became a dictatorship. Germany became a police state hungry for territory and racial supremacy. Those members of German society deemed "impure" or "unworthy" by the Nazis, such as Jews, gypsies, the handicapped, Communists, and homosexuals, were increasingly persecuted during the 1930s. When Hitler launched the September 1, 1939, attack on Poland that started World War II, the path was made for the mass killings of the Holocaust. Approximately 11 million people were killed as a result of

---

John Lukacs, *The Hitler of History*. New York: Vintage Books, 1997. Copyright © 1997 by Random House. Reproduced by permission.

Hitler's genocidal policies; 6 million of them were Jews.

With Germany's defeat in sight, Hitler committed suicide in a bunker on April 30, 1945. The war in Europe ended several days later, but the war Hitler started for Germany's "living space" would continue in the Pacific theater until August 10, 1945. About 53 million soldiers and civilians perished as a result of World War II.

In the following excerpt John Lukacs analyzes Hitler's role in history. He examines the consequences of Hitler's rule for both Germany and the world. Lukacs investigates, too, the way Hitler's vision affected political movements both in his own day and during the rest of the twentieth century. He assesses Hitler's place in the twentieth century and evaluates both the long- and short-term impact of his ideas and actions.

Historian John Lukacs has written more than twenty books on twentieth-century European history, particularly World War II and related topics. He was a professor of history at Chestnut Hill College in Philadelphia until his retirement in 1994.

"There is no argument," [English writer] George Orwell once wrote, "by which one can defend a poem. It defends itself by surviving, or it is indefensible." So we judge achievements by their consequences. And what were the consequences of Hitler?

The Consequences of Hitler

His war (and World War II was *his* war) ended with the greatest catastrophe for the German and also for the Central and Eastern European peoples—for the latter, because the Russian occupation and the subsequent Communist rule in that part of Europe was the consequence of that war. A consequence of the war was the division of Germany, which lasted for more than forty years, with mass expulsions from the east and protracted sufferings for the German people in the so-called "German Democratic Republic." A more enduring consequence was the drastic reduction of Germany in the east: the entire loss of East Prussia, together with other substantial portions of the former Prussia, Silesia, Saxony to Poland (and, in the case of the Königsberg enclave, to Russia). Even more enduring: the almost complete elimination of the presence of ethnic Germans from countries in Eastern Europe where some of their ancestors had lived for eight hundred years.

The year 1945 marked, too, the end of the predominance—political, cultural, intellectual—of Europe in the world, the end of the European age, and the end of the European state system. There was something else, too: the end of the predominance of German intellectual influence in the world. For what had happened, for about seventy years after 1870, was not only that the practices and standards of German education and learning had influenced and were adopted in many places of Europe and of the world, including nations that were opponents of Germany in the two world wars. Something that had begun after the French Revolution: a romantic (and often sentimental and categorical) idealism, reacting against the materialism (and often against the rationalism) of the Enlightenment. This most important and potentially fruitful intellectual achievement—a great chapter in the history of the European mind—was mostly represented and exemplified by Germans; and then it was carried by some of them to extremes, to a deterministic idealism that proved to be more inhumane than the deterministic materialism that had preceded and (lamentably) survived it—at least for a while. And an incarnation of an unstinting belief in a determinist idealism was Adolf Hitler.

Hitler's Influence on National Socialism

The English historian of religion Owen Chadwick wrote: "The Reformation would have happened without [German religious reformer Martin] Luther. But without Luther it would not have happened in the way it happened." Four centuries later the same condition—mutatis mutandis ["with the necessary changes"]—applies to nationalist socialism and Hitler. He was the greatest revolutionary of the twentieth century. That is not an approbatory adjective. "Great" may be applied to criminals or terrorists, too. (Also, one need not be a reactionary to sense, especially at the end of the twentieth century, that to designate someone as a Great Conservative may be even more approbatory—at least in the Western world—than the epithet of a Great Revolutionary.) In any event, what followed Hitler was our still-present era of no great wars and no great revolutions—something that he had not foreseen. . . .

What he had seen—and, more or less, accurately—was the formidable attraction of populist nationalism in the age of the masses. That nationalism proved to be the principal political reality in the twentieth century. He was its most extreme representative. He sensed that sometime after 1870 nationalism and socialism came to supersede the older nineteenth-century categories of conservatism and liberalism, indeed perhaps even of Right and Left; and that, of the two, nationalism was more influential than socialism. The categories of socialism and capitalism themselves began to be outdated, because strength was more powerful than wealth, because nationality was more powerful than class, because nationalism was more powerful than internationalism. When there was national unity, the formerly rigid categories of socialist and capitalist, public and private ownership began to leak; what mattered was not ownership than management; and ownership and management and labor would be ultimately obedient to the dictates of nationalism.

Hitler's Vision

Hitler was not the founder of National Socialism, not even in Germany. Among the principal figures of the century, [Italian dictator Benito] Mussolini was the first national socialist who recognized, around 1911, that he was an Italian first and a Socialist second—this eight years before his creation of "Fascism." We have seen differences, not only in the practices but in the ideas, of Hitler's National Socialism and Mussolini's Fascism. But that becomes inconsequential when we look at the reciprocal influences of the main dictators in the 1930s and 1940s: Mussolini, [Soviet leader Joseph] Stalin, and Hitler. That comparison of reciprocal influences—more precisely, of their development—ought to tell us something. Mussolini was not (and did not become) a Communist; Stalin was not (and did not become) a Fascist; but both of them became influenced and impressed by Hitler's ideas and achievements to an extent that makes it both proper and precise to say that they became more and more nationalist socialists as time went on. That development (including their increasing inclination to anti-Semitism) ended with their deaths. So much for the war and the short run. But we must recognize that in the long run, too, in one sense Hitler's vision survived him. During the twentieth century the compound of nationalism with socialism has become the nearly universal practice for all states in the world. International socialism is a mirage. At the same time every state in the world has become a welfare state of sorts. Whether they call themselves socialist or not does not matter much. Hitler knew that. The economic structure of Germany that he had in mind had few of the characteristics of either Marxian or state socialism, but it could not be called capitalist, either. Fifty years later it cannot be denied that nationalism remains the most potent force in the world. We are all national socialists now. Of course, the proportions of the compound of nationalism and socialism vary from country to country; but the compound is there, and even where social democracy prevails it is the national feeling of people that matters. What was defeated in 1945, together with Hitler, was German National Socialism: a cruel and extreme version of nationalist socialism. Elsewhere nationalism and socialism were brought together, reconciled and then compounded, without remotely comparable violence, hatred, or war. But Hitler's nationalism was profoundly different from traditional patriotism, just as his socialism had few of the marks of the traditional philanthropy of the earlier Socialists.

To this the objection may be raised: After all, has not the appeal of Communism long survived that of Hitler's National Socialism? Despite all the superficial evidence—examples: the surviving appeal of "Communist" parties in Russia and Eastern Europe, Chinese "Communism," [dictator Fidel] Castro in Cuba—the answer is no, for three reasons at least. The first is that the surge of Communism that enveloped much of Eastern Europe after 1945 was not the result of popular revolutions but simply due to the presence of Russian armed forces in that part of Europe. The second is that the sporadic rise of Communist regimes in the oddest places of the so-called Third World—Cuba, Ethiopia, Angola, among others—was the obvious result of anticolonialism (and in Castro's case, of anti-Americanism) rather than of the appeal of Communism as such or the example of the Soviet Union. The third reason, connected with the second, is that the present—and probably transitory—reappearance of Communist or pro-Stalinist parties, especially in Russia, is not only inseparable from but fundamentally bound to a resurgent and populist nationalism. If International Socialism is a mirage, International Communism does not even qualify as an optical illusion.

As Karl-Dietrich Bracher put it: "The kernel of the phenomenon of Hitler was a fundamental underestimation of [the attraction of] National Socialism"—that is, not only of Hitler but of the idea he seemed to represent. And in this respect it behooves us to consider the corresponding, and perhaps ominous, statement by another German historian, Hagen Schulze . . . about that important experience "in our history: that the constitutionalists of the first German republic had nothing effective to counter the enormous emotional appeal of the nationalists. Certainly the experience of the Third Reich has considerably dampened the German inclination to nationalistic extremes," but it is questionable "whether this kind of dampening will last more than one or two generations, despite all of the political pedagogy, about the efficacy of which one should have no illusions." Neither should we have illusions about the permanent constitutional validity of laws according to which in Germany the public display of the swastika and of pictures of Hitler remains forbidden forever. What we must hope for and trust is that when the time for the removal of such proscriptions comes, such a legal decision will reflect a climate in which the symbols of Hitler's era will attract nothing more than historical curiosity.

Hitler's Place in the Twentieth Century

That time is not yet; and now when the German state has become united, when the entire postwar period marked by fifty years of the so-called cold war has unraveled, when the entire century is ending, let us conclude with an attempt to identify—identify, rather than define—Hitler's place in the history of it.

The twentieth century—historically speaking—was a short century. Whereas the historical eighteenth century lasted 126 years, (from 1688 to 1815), marked by the world wars principally between England and France; and the nineteenth century lasted ninety-nine years (from 1815 to 1914), marked by the absence of world wars, the twentieth century lasted seventy-five years (from 1914 to 1989), marked by the two world wars and by their consequence—the so-called cold war between America and Russia. It ended in 1989, with the withdrawal of Russia from Eastern Europe and the reunification of Germany. The crucial period of this—transitional—century (marking the passage from the so-called Modern Age into something else) was its early one-third (1920 to 1945) in which, of course, the career of Hitler belongs. During this period—again contrary to the popular and mistaken impression about the importance of the Russian Revolution in 1917—the history of the world (and not only of Germany or of Europe) was marked by the existence and the competition of a triangle of forces. There was parliamentary democracy, incarnated principally by the English-speaking nations, by the states of Western Europe, and Scandinavia. There was Communism, represented only by Soviet Russia, incapable of assuming power elsewhere. And there was National Socialism (and also other, to some extent similar but by no means identical, nationalist movements) in Germany, after 1933 incarnated by Hitler and the Third Reich, which proved to be so powerful that it took the unnatural and temporary alliance of Liberal Democracy and Russian Communism, of the English-speaking and the Russian empires, to defeat and conquer it. Neither side could do it alone.

The Hitler Phenomenon

That by itself identifies Hitler's place within the history of the twentieth century. But there was even more to it. This triangle, represented by liberal-conservative-democratic people, and by Communists and radical nationalists, repeated itself in every country in Europe, and in almost every country of the world—including the United States, South America, China, and Japan. (In 1945, the radical nationalists were silenced or subdued—temporarily. Later their successors reappeared again.) Not all of the radical nationalists were followers or admirers of Hitler, though many of them were. [German historian Percy] Schramm wrote that we cannot ignore the fact that Hitler represents a key phenomenon in the history of Germany in the twentieth century, but not only in the history of Germany.

The German term "Ortsbestimmung" is defined as "position-finding" in a dictionary sense; but it is a word often used by German historians, meaning the definition of a place in history, and suggesting that while historical judgments of an event or person may vary, the *place* of the events or persons within the sequence of history is—eventually—ascertainable. In this respect [journalist Barnhard] Schreiber's relatively recent conclusion is still valid: "The place of the National Socialist period in history still remains a desideratum of further research." But is the problem still that of "research"? Perhaps—even though there is reason to believe that no more important, or even significant, documents by or about Hitler will come to light. But then history does not only consist of documents.

Hitler Gives History a New Direction

by Joseph Goebbels

To many of his followers and admirers, Adolf Hitler was regarded as a man who changed history for the better. His birth on April 20, 1889, was considered a special moment in German history; as a result, his birthday was deemed a holiday in Nazi Germany.

Joseph Goebbels was minister of popular enlightenment and propaganda in Nazi Germany from 1933 to 1945. As a devoted follower of Hitler, he was responsible for "educating the public" by controlling the press, censoring works by Jews, coordinating mass rallies, and other means. One of Goebbels's yearly projects was to write and deliver speeches on the radio in commemoration of the Nazi leader's birthday.

Goebbels delivered the following speech on April 19, 1939, the evening before Hitler's fiftieth birthday. He states that Hitler has bestowed his name on a historical era and then recounts the changes in Germany since the führer came to power. Calling Hitler a man of historical greatness, Goebbels underscores the uniqueness of both Hitler's vision and his leadership qualities.

I n an unsettled and confused world, Germany tomorrow celebrates a national holiday in the truest sense of the word. It is a holiday for the entire nation. The German people celebrate the day entirely as a matter of the heart, not of the understanding.

Tomorrow the Führer finishes his fiftieth year. The entire German nation takes pride in this day, a pride in which those peo-

---

Joseph Goebbels, speech honoring Hitler's fiftieth birthday, April 19, 1939.

ples who are friendly with us also take deep and hearty part. Even those who are neutral or oppose us cannot ignore the strong impact of the events. Adolf Hitler's name is a political program for the entire world. He is almost a legend. His name is a dividing line. No one on earth can remain indifferent to his name. For some, he represents hope, faith and the future, for others he is an exemplar of confused hatred, base lies and cowardly slander.

The highest that a person can achieve is to give his name to an historical era, to stamp his personality indelibly on his age. Certainly the Führer has done that. One cannot imagine today's world without him.

[The nineteenth-century German historian Heinrich von] Treitschke once said that men make history. If this is true, when more so than in our era? He has shown his simplicity and depth in the most wonderful way. Adolf Hitler has influenced not only the historical development of his country, but one can say without fear of exaggeration that he has given all of European history a new direction, that he is the towering guarantee of a new order for Europe.

Our part of the world looks vastly different today than it would without him, not to mention his impact on our own people and nation. He has given the German nation an entirely new face through revolutionary internal transformations.

Someone who saw Germany for the last time in 1918 [after Germany's defeat in World War I] would scarcely recognize it today. The people and nation are entirely different. What seemed like a miracle only a short while ago is self-evident today.

About a year ago, the Führer solved the problem of joining Austria to the Reich. The whole people celebrated his 49th birthday then. 7½ million Germans had returned to the Reich. A Central European problem one almost believed to be unsolvable was miraculously solved.

On the eve of his 50th birthday, we can happily see that once again the map of Europe has changed in the Reich's favor, and—unique in world history—this change has occurred without bloodshed. It came as the result of a clear desire to establish peace in an area of Europe in which the contradictions were so severe that there was danger that they sooner or later would cause a general European conflagration.

This new peace in the threatened areas is not a peace of tired, moralistic theories that are endangered as often as the false bourgeois democrats praise them. It is much more a peace that is built on practical realities.

Such a peace could be built only on the foundations of a higher, instinctive understanding growing from the knowledge that only strength gives a people the opportunity to finally resolve problems.

Successful policies require both fantasy and reality. Fantasy as such is constructive. It alone provides the strength for powerful, flexible historical conceptions. Realism on the other hand brings the ideas of political fantasy in agreement with hard reality.

A Man of Historical Greatness

The Führer possesses both characteristics in a unique harmony seldom seen in history. Fantasy and reality join in him to determine the goals and methods of political policy. His contemporaries are constantly astonished and amazed by seeing how he brilliantly brings goals and methods together to influence history. He has no stubborn ideas, no tired tactical doctrines to dim his vision and reduce his political imagination. His inflexible principles are joined with changing and flexible political methods that have led to the greatest and most unexpected successes for Germany.

That is nothing new for us old National Socialists. We learned to admire the Führer's political abilities in the earliest phases of our party's hard struggle for power in the Reich. They were demonstrated in many small and apparently unimportant ways at the time, though they were then for us and the movement as important as the goals and problems of today.

Then too there were doubters who failed to see the greatness and brilliance of the Führer's decisions during the struggle for power. They favored the false wisdom that [Prussian army officer and writer Carl von] Clausewitz discussed: they wanted nothing but to escape danger. We are therefore not surprised or anxious to see the same or similar happenings in internal German politics that we earlier saw in the National Socialist movement.

The only thing that has changed over the years is the scale of the Führer's actions; his methods and goals have remained the same. Back then we saw in him the political instincts of a truly historic genius, able to understand problems and find the simplest and clearest solution to them from his own greatness and certainty. That is why we were then the most loyal and obedient servants of this man and his work, entirely aside from the human element.

So what we see today is nothing new for we old National Socialists. We therefore have no doubt of the outcome of Germany's current battle for its national existence. Our whole people have the same instinctive feelings, which are the cause of the blind and unshakable confidence they place in the Führer.

The man in the street is usually not in a position to understand the entire political situation. He lacks the practice, the experience

Hitler's followers admired his unique vision and believed that he changed history for the better.

and above all the background necessary to form a clear and certain judgment. It is therefore entirely understandable why he dislikes theories and programs, and prefers to place his firm and confident faith in a personality.

A nation inclines to doctrines only when it is poor in personalities. But when a man of historical greatness stands at their head, one who not only wants to lead but is able to do so, the people will follow him with their whole hearts, giving him their willing and obedient allegiance. Even more, they will put all of their love and their blind confidence behind him and his world.

Hitler's Birthday Is a German Holiday

A nation is willing to sacrifice when it knows what it is sacrificing for and why it is necessary. That is true in Germany today. None of the numerous slogans that the broad masses of our people heard in the years after 1918 has had such powerful effect on the entire nation as the phrase "One People, one Reich, one Führer!"

The first two phrases were heard for the first time in 1937 at a singing festival in Breslau. The Führer stood high on the platform against the gathering darkness. Hundreds of thousands of people had gathered from every corner of the nation and from everywhere in Europe where Germans dwell to hear him speak. Suddenly, from the corner of this army of hundreds of thousands where the Austrians stood came the call "One people, one Reich." It gripped and fascinated the whole crowd, and for the first time gave concise but clear expression to a program.

A year later we saw the Führer on a hot Sunday afternoon standing on the platform at the Schlossplatz in Breslau once again. German gymnasts performed before him. As the racial comrades from the Sudetenland passed before him, without command or order, they suddenly formed a wall before him. These people who had come from the Sudetenland to Breslau only to see his face, refused to move. Weeping women seized his hand. One could not understand what they were trying to say, since tears drowned their voices.

Once again, it was only a few months before the problem they had brought to the Führer was solved.

The Greater German Reich, in the truest sense of the word, has now become a reality. Even more, the Führer has given his peace to Central Europe. It is clear that this is not to the pleasure of those democratic enviers of the National Socialist Reich. Through the Treaty of Versailles they had built a ring of trouble spots around Germany that they could use to keep the Reich in constant difficulties.

A man has come from the broad masses of the German people who removed these trouble spots with the firmest measures. Democracy sees its hopes vanishing. That explains their rage and moralistic disappointments. Their hypocritical prayers came too late. The enemies of the Reich are at the end of their rope. They look ridiculous, and cannot understand why.

We greet their hysterical cries with sovereign contempt, a sovereign contempt shared by the entire German people. The German people know that the Führer has restored them to their deserved position in the world. The Reich stands in the shadow of the German sword. Germany's economy, culture and popular life are blooming in a security guaranteed by the army. The nation, once sunk into impotence, has risen to new greatness.

We remember all of this as we begin to celebrate the 50th birthday of the man whom we thank for our nation's might and our people's greatness. No German at home or anywhere else in the world can fail to take the deepest and heartiest pleasure in participation. It is a holiday of the nation, and we want to celebrate it as such.

A people fighting for its fate must now and again stop in the midst of the tumult of events to remind itself of its situation, methods and goals. Today is such a time. The nation puts on its best clothing and stands before its Führer united in loyalty and brotherhood, to bring him their heartiest best wishes on his 50th birthday. These are the wishes of all Germans in the Reich, as well as those in every other nation and continent. Germans throughout the world join with us who have the good fortune to live in the Reich in these warm and thankful wishes. To this choir of a hundred million are joined the voices of all those peoples who want true peace and order in Europe, who love its history and its culture.

As we begin to celebrate the Führer's 50th birthday in this festive hour as a great national community, we join in a fervent prayer to Almighty God that he graciously preserve in the future his life and work. May he grant the German people's deepest wish and keep the Führer in health and strength for many more years and decades. Then we will not need to fear for the future of the Reich. The fate of the German nation rests in a strong and sure hand.

We, the Führer's oldest followers and fellow fighters join together at this festive hour with the hearty wish that we have always had on the birthday of this man: May he remain for us what he is and always was:

Our Hitler!

The Significance of l'Affaire

by Gordon Wright

The Dreyfus affair began in 1894 when a French spy found a list of secret military documents in the wastebasket of the German military attaché in Paris. French army officer Alfred Dreyfus, a Jew, was accused and arrested for this treasonous act despite the fact that no evidence implicated him in the crime. Dreyfus proclaimed innocence but was found guilty and sentenced to exile on Devil's Island. His family tried unsuccessfully to reopen the case. It was not until 1896, when a new counterintelligence chief found evidence suggesting that another officer, and not Dreyfus, was likely guilty, that the matter resurfaced. In August 1899 Dreyfus was retried and again found guilty. However, he was soon pardoned.

In this selection Gordon Wright provides a synopsis of the events that unfolded in the final years of the nineteenth century. He also explains how public opinion erupted and why France became divided over the Dreyfus case. The reinvestigation of the issue was covered in the mass circulation daily newspapers, and journalists began to express either pro- or anti-Dreyfus sentiments. Wright notes that both the educated elite and the general public took sides, and debates over Dreyfus's guilt or innocence became politically tied to questions about justice, anti-Semitism, and the interests of the state.

Gordon Wright was a professor at Stanford University and a prominent scholar of modern France. He was one of America's most accomplished historians, with both a prolific writing career and a fine reputation as a teacher.

Gordon Wright, *France in Modern Times: From the Enlightenment to the Present.* New York: W.W. Norton & Company, 1981.

F rom 1896 until the end of the century France was sucked into the vortex of an almost unprecedented crisis—a crisis that deserves the name of a bloodless but no less intense civil war. This was the Dreyfus case; or, in French usage, simply *l'Affaire.*

A Case of Alleged Treason

Few fictional thrillers are more complex than this real-life case of alleged treason. So many of the basic facts remain in dispute that even a summary of the actual events is open to challenge. In 1894 an unsigned letter apparently destined for the German military attaché in Paris (and probably filched from the attaché's wastebasket) came into the possession of the French army's counter-intelligence unit; it listed several secret military documents which the author was prepared to deliver to the Germans. After a brief investigation, suspicion fell on a French staff officer, Captain Alfred Dreyfus. Arrested by the army and privately pressed to confess, Dreyfus indignantly asserted his innocence. Handwriting experts failed to agree as to whether Dreyfus had actually written the incriminating letter, and army investigators failed to uncover any persuasive supporting evidence. The case might have been dropped had not that professional anti-Semite, Edouard Drumont, heard rumors of the affair and raised a public outcry. Drumont (whose best-selling book *Jewish France* [1886] has been called "the first great explosion of modern anti-Semitism") charged that wealthy Jews were trying to buy the freedom of a traitor who had been caught red-handed. Army authorities decided, therefore, to proceed with a court-martial from which the public was excluded. The verdict of guilty, it later turned out, rested heavily on a secret dossier presented to the court but never shown to Dreyfus or his attorney. Dreyfus was sent to Devil's Island for life, while Frenchmen applauded the army's efficiency and toughness. The Socialist Jean Jaurès even commented in the Chamber that if Dreyfus had not been a bourgeois and an officer, he would have been shot. Efforts by the Dreyfus family to get the case reopened aroused no support.

Military secrets nevertheless continued to flow to the German Embassy. In 1896, two years after the original trial, the new chief of counter-intelligence, Colonel Picquart, reexamined the Dreyfus file for leads to possible collaborators in treason. Astonished and dismayed by what he found, Picquart concluded that Dreyfus had not written the original letter, and his suspicions turned to another officer, Major Esterhazy, known to be of dubious character and habits. Picquart reported his findings to his superiors, and urged that the Dreyfus case be reopened; when he persisted, they transferred him to a post in the Tunisian desert. But something of what he knew began to leak out to the press, and a few newspapers (notably Georges Clemenceau's) launched a crusade for a new trial. Meanwhile another counter-intelligence officer, Major Henry, fed bits of anti-Dreyfus information to Drumont and other journalists, so that the controversy grew increasingly public and bitter. Looking to the future, Henry also took the precaution of forging some new documents for the Dreyfus file in order to strengthen the case against him. Army authorities steadily resisted all demands for a reopening of the case—the more so because they resented civilian interference in a matter of military justice. At last, in an effort to end the matter once and for all, the army staged a court-martial of Major Esterhazy (January 1898); he was triumphantly acquitted. Two days later the novelist Emile Zola published in Clemenceau's newspaper a long article headed *"J'accuse,"* openly charging the army's top commanders with conniving at deliberate injustice. Zola was promptly arrested, tried, and sentenced to a year's imprisonment. Rather than serve his term, he departed into English exile.

Taking Sides

By this time (1898) virtually the entire educated elite of France was committed to one side or the other. Families were split; there were frequent clashes in the streets, and duels were fought daily. Most intellectuals became vociferous Dreyfusards; but a smaller group of intellectuals joined the monarchists and right-wing Catholics to resist a new trial for Dreyfus. The Assumptionist organ *La Croix* leaped into the fray and campaigned for stripping all Jews of citizenship. [The Assumptionists are a Catholic order of priests and brothers.] Most politicians still tried to quiet the storm and to stay aloof from any commitment; but there were notable exceptions like Clemenceau, Jaurès, and the Alsatian Senator Auguste Scheurer-Kestner who held that party interests should give way to the overriding principle of justice for the individual citizen. Even Jaurès, that dedicated apostle of human rights, had been slow to come out for Dreyfus; many of his Socialist colleagues remained reluctant until the end.

A major break in the case came suddenly in August 1898, when a new counter-intelligence officer accidentally detected Major Henry's forgeries in the Dreyfus file. Henry confessed, and a day later cut his throat in his jail cell. Esterhazy fled to England, where he gave out conflicting stories about his role in the case. The anti-Dreyfus faction still insisted, however, that Dreyfus was the culprit, and that there were top-secret documents in the files of counter-intelligence to prove it if they could only be made public without causing a war. The army could no longer resist a new trial. In August 1899 Dreyfus was brought back from Devil's Island to face a second court-martial, public this time. In a curious verdict, the military judges found him guilty, but with extenuating circumstances. He was promptly pardoned by the president of the republic and was fully vindicated by a civilian court several years later.

There are enough gaps and contradictions in the evidence to leave the affair still partially shrouded in uncertainty. Only a few Frenchmen still believe that Dreyfus was guilty and was saved by Jewish influence and money. Some think that there may have been two traitors—Esterhazy and a second officer who, to avoid discovery, continually fed in false leads to distract the pursuers. Still others suggest that there may have been no traitor at all; that Esterhazy, for somewhat obscure reasons, had been ordered by his superiors to write the note of 1894 to the German attaché. The list of hypotheses is almost endless.

Political Effects of the Dreyfus Affair

More important, however, was the impact of *l'Affaire* on French political life and the enduring myth that it left behind. Its immediate effect was to bring the Radicals into power and to keep them there almost steadily until the Great War. The Moderates were badly split over the Dreyfus issue and were mainly concerned with keeping the lid on; the Radicals profited by the opportunity to come out on the winning side when victory for the Dreyfusards had become almost certain. They convinced themselves, and a great many voters, that they had saved the republic from a clerico-monarchist plot, and that they must be kept in office to protect the republic. The new Waldeck-Rousseau cabinet (1899–1902), though headed by a Moderate, was dominated by Radicals and supported by Socialists; it was a kind of popular front ahead of its time, cemented by the presumed right-wing threat to the regime.

L'Affaire also sharpened the suspicions of French leftists toward both the army and the church. A new wave of antimilitarism and anticlericalism followed at once; steps were taken to republicanize the officer corps and to weaken church influence. Both the general staff and certain elements in the church had blundered badly in their handling of the Dreyfus controversy; their behavior naturally fed the republicans' suspicions. The general effect was to deepen existing divisions within France and to set back the growth of something like national consensus.

Significance and Myth

Like most great emotional upheavals, this one was significant mainly for the mythology it inspired. Most French republicans saw the Affair as a contest between the forces of righteousness and progress on the one hand and those of bigotry and obscurantism on the other. The Dreyfus camp asserted the principle of justice for every man, even though the institutional foundations of society and the state might be temporarily shaken. The anti-Dreyfus factions stood for the principle of *raison d'état*, in which the end justifies the means; they rejected the idea that the individual citizen possesses any sacred rights transcending those of the state. To some degree, the controversy *was* rooted in this fundamental conflict in values, and the victory of the Dreyfusards did reinforce a certain view of human rights. But the case was also heavily encrusted with less admirable motives of political advantage and bigoted loyalty; and in large part the story was one of fallible human beings sucked into the maelstrom without quite knowing how they got there. One curious by-product was the strengthened self-image, and popular image, of intellectuals as incorruptible men of virtue, unsullied by prejudice and able always to detect the right and just cause. Along with it went the legend of honesty and fairness as characteristics of the left, and selfishness and bigotry as qualities of the right. Both illusions were of strictly limited validity and were to produce some unfortunate consequences in twentieth-century French political life. Thus the complexities of the case itself were hardly greater than the complexities of the heritage left by it. One might epitomize *l'Affaire* as "the shame and the glory of modern France." And one might ask whether a Dreyfus Affair could have occurred anywhere except in France.

8

The Dreyfus Affair Challenges France:
October 15, 1894

The Injustice of the Dreyfus Affair

Émile Zola

Émile Zola was a French writer. In addition to his successful career as a novelist, Zola played a public role in the Dreyfus affair. He was strongly convinced that French army officer Alfred Dreyfus was wrongly accused of treason and that Dreyfus's conviction in 1894 was an injustice. The officer's brother, Mathieu Dreyfus, as well as lawyer Louis Leblois and French Senate vice president Auguste Scheurer-Kestner, sought to rehabilitate Dreyfus and appealed to Zola to become involved in challenging the case.

Zola's campaign to call attention to the unfairness of the case consisted of publishing open letters, such as the following "Letter to France," in newspapers. Written January 7, 1898, this letter is an appeal to the reason of the French people. Because Dreyfus was Jewish and anti-Semitism was prevalent in French society at the close of the nineteenth century, Zola urges the people of France to reject anti-Jewish sentiment and to seek justice instead. He also cautions the public about the press and warns the French not to believe everything presented in the newspapers about the Dreyfus case. The French people should think for themselves, he maintains. Zola expresses hope that France will unite and demand the truth about the Dreyfus affair in order to free an innocent man.

Zola's letters ultimately led to the reopening of the case, and Dreyfus was tried again in 1899. Initially Dreyfus was found guilty and was then pardoned, but eventually the verdict was reversed. Zola, for his part in challenging the authorities, was accused of libel

---

Émile Zola, "Letter to France," January 7, 1898.

and was sentenced to imprisonment. He fled to England, though, and returned to France after Dreyfus was cleared of all wrongdoing.

We are going through a frightful time of moral confusion; the public's conscience appears to be clouding over. At this time, France, it is to you that I must speak, to the nation, to the mother country.

The People of France Must Think for Themselves

Every morning, France, when I read in the papers what you seem to think of this lamentable Dreyfus Affair, my stupefaction increases, my reason rebels still more. Is it possible that you, France, have come to this? Have you really been convinced by the most blatant lies? Are you really siding with the evildoers and their muck, against a few decent individuals? Are you really losing your head on the asinine pretext that your army has been insulted and there is a plot to sell you to the enemy? Why, the wisest and most loyal of your children have in fact but one desire: that, as all Europe turns its eyes attentively towards you, you should remain the nation of honour, of humanity, truth and justice.

Indeed, the general public has come to that, especially the little people, the humblest people in the towns and almost everywhere in the provinces and in the country, that great majority of the people who accept what their newspapers or their neighbours tell them, who have no way of gathering information for themselves, no way of stopping to think for themselves. France, what has happened? How have your goodhearted, common-sensical people let fear make them so ferocious and intolerance plunge them into such darkness? Your people are told that a man who may be innocent is being tortured in the most appalling way: there is moral and material evidence that a revision of his trial is indispensable—and yet your people flatly refuse to see the light, they hide behind the scoundrels and the sectarians who find it in their interest to let the body stay buried. Are these the same people who not long ago would have torn down the Bastille all over again to rescue a single prisoner from it?

What anguish, what sadness in the souls of all those who love you, France, who strive for your honour and your grandeur! With what distress I behold these rough and stormy seas and try to discern the causes of the tempest that threatens to make your greatest glory founder. Nothing could be more deadly serious; no symptoms could be more alarming. I will dare to speak out, for all my life I have had but one passion: the truth, and I am merely continuing my life's work.

Don't you realize that the danger lies precisely in the wilful obscurantism of public opinion? Day after day a hundred different newspapers repeat again and again that public opinion does not want Dreyfus to be innocent, that his guilt is essential to the country's salvation. But France, don't you realize how very guilty you would be if your leaders allowed themselves to use such a sophism to stifle the truth? It would be the fault of France itself for having insisted on finding a crime, and then what a responsibility France would bear one day! That is why those of your sons who love you and honour you, France, have but one ardent duty in this gravest hour: to act powerfully on public opinion; to enlighten it and rescue it from the error towards which blind passions are steering it. There is no more useful, more sacrosanct task than this.

Lies in the Press

Oh, I'll speak to them, all right—with all my might, I'll speak to the little people, the humble people who are being poisoned and forced into delirium. That and that alone is the mission I assign myself. I will cry out to them, tell them where the country's soul and its invincible energy and its undoubted triumph really spring from.

Let's see where things stand. A new step has just been taken: Major Esterhazy has been summoned before a court martial. As I have been saying from the very first day, the truth is on the march and nothing shall stop it. Despite the attempts to stand in its way, every step forward will be taken, mathematically, in due time. The truth carries a power within it that sweeps away all obstacles. And whenever its way is barred, whenever someone does succeed in burying it for any time at all, it builds up underground, gathering such explosive violence that the day it bursts out at last it blows up everything with it. Just try to keep it walled up a few months longer behind lies or behind closed doors, and ultimately you'll see that you have paved the way for the most shattering disaster.

But as truth moves forward the lies pile up, denying that truth is on the march. Nothing could be more significant. When General de Pellieux, in charge of the preliminary inquiry, submitted his report which came to the conclusion that Major Esterhazy might be guilty, the foul press made up the following story: that at Esterhazy's express wish, General Saussier—hesitating, convinced of Esterhazy's innocence—agreed to turn him over to military justice, in order to keep him happy. Today the papers are doing better still. They are saying that since three experts have again recognized the bordereau [list of documents] as having been beyond any doubt written by Dreyfus, Major Ravary, in the course of his investigation, concluded that the case must be dismissed and that the reason why Major Esterhazy is to appear before a court martial is that he had once again forced General Saussier's hand, demanding to be judged after all.

Isn't all this intensely comical! Isn't it all utterly stupid! Can you picture this defendant leading the whole case, dictating the decrees? Can you believe that whereas two investigations have concluded that the man is innocent, they are now going to the tremendous trouble of holding a court session for his benefit, for the sole purpose of producing a decorative comedy, a sort of judicial apotheosis? This would merely be a mockery of justice, since they declare that he is certain to be acquitted; after all, justice is not intended to judge the innocent, and the least we can ask is that the court's ruling will not be written off-stage, in the wings, before the session even begins. Since Major Esterhazy is summoned before a court martial, let us hope for the sake of our nation's honour that this is a serious business and not just some sham to keep the idle onlookers amused. Poor, poor France, they must take you for a fool, if they try to palm off such nonsense on you.

And similarly, those reports that the foul press publishes—aren't they all a heap of lies? They should be enough to open your eyes. Personally, I flatly refuse to believe that those three experts failed to recognize, at their very first glance, that Major Esterhazy's handwriting and the writing of the bordereau were absolutely identical. Take any small child going by in the street, bring him upstairs, place both pieces of writing in front of him and he will tell you. 'These two pages were written by the same gentleman.' You don't need any experts; anyone will do; the resemblance between certain words is that obvious. So true is it, in fact, that the Major has recognized this frightening resemblance and, to explain it away, he claims that tracings were made of sev-

eral of his letters and launches into a whole laboriously complicated and thoroughly childish story that the press has been busy
with for weeks. And yet they come and tell us that three experts
have been found who declare once again that the bordereau is indeed in Dreyfus's writing! No no, it's too much! Such audacity
becomes suspect in the end, and decent people are going to get
angry, I hope!

The Basis of Public Opinion

Some papers go so far as to say that the bordereau will be omitted,
that it won't even be mentioned in court. Then what *will* be mentioned? and why will the court bother to sit? That's the crux of the
Affair: if Dreyfus was sentenced on the strength of a piece of evidence written by another person and if that piece of evidence is
enough to prove that other person guilty, then a revision of the trial
is indispensable—the logic requiring it is undeniable, for it is not
possible to declare two people guilty of the same crime. Maître
Demange has repeated, officially, that the only piece of evidence
communicated to him was the bordereau; Dreyfus was legally
found guilty only on the strength of that bordereau, and, even supposing that in defiance of all the rules of legal proceedings there
are pieces of evidence that have been kept secret (personally, I cannot believe that), who would dare to refuse a revision once it was
proven that the bordereau—the only known, acknowledged piece
of evidence—is in another man's handwriting? And that is why
there has been such an accumulation of lies concerning that bordereau, for it is at the very heart of the entire Affair.

Here, then, is the first point to be noted: to a large extent public opinion is arrived at on the basis of those lies, those idiotic
and extraordinary tales the press prints every morning. The time
will come when responsibilities will have to be taken and accounts will have to be settled with the foul press that dishonours
us in the eyes of the whole world. Some of those newspapers are
perfectly at home in that role; they've never done anything else
but sling mud. But how astonishing and sad it is to find that
among them is *L'Echo de Paris*, a literary sheet that is so often
in the vanguard of ideas and yet here it is playing such a nasty
role in this Dreyfus Affair! Its articles are scandalously violent
and prejudiced—and unsigned. It is said that the people behind
them are the very ones who were so disastrously clumsy as to
have Dreyfus found guilty. Does M. Valentin Simond realize that

they cover his paper with shame? And there is another paper whose attitude should smite the conscience of all decent people; I refer to *Le Petit Journal*. If the insidious broadsheets that have a circulation of only a few thousand howl and lie in order to boost their circulation, that is understandable, and besides it does only limited harm. But when *Le Petit Journal*, with a circulation of over one million, which speaks to the ordinary people and reaches everywhere, disseminates error and leads public opinion astray, then matters are exceptionally grave. When a newspaper has such a moral responsibility, when it is the spiritual leader of such a large flock, it must be of the most scrupulous intellectual integrity lest it commit crimes against the public good.

And that, France, is the first thing I find in the madness that is sweeping you away: the lies in the press, the diet of inept farce, low insults and moral depravity that it feeds you every morning. How could you possibly demand truth and justice when they are doing so much to denature your legendary virtues, the clarity of your intelligence and the sturdiness of your reasoning?

The Poor Behavior of France

But there are facts that are more serious still, a whole set of symptoms which turn the crisis you are going through into a terrifying lesson, for anyone who knows how to see and judge. The Dreyfus Affair is merely a deplorable incident. The terrible thing that must be admitted is the way you are behaving throughout this business. You look healthy enough—but suddenly little blotches appear on the skin: death is there, inside you. All the political and social poison you have absorbed suddenly breaks out on your face.

Why did you let them get away with it when they shouted that your army was being insulted? (and why did you ultimately shout it too?) whereas all that the ardent patriots wanted was to defend the army's dignity and honour. What is your army, after all? Why, today, it is you, it is all of France. The army is not just a certain chief, not just a certain officers' corps, not just the bemedalled higher ranks. The army is all of your children, ready to defend France's soil. Ask yourself frankly: was it really your army you were rushing to defend when no one was attacking it? What you suddenly needed to cheer on was the sabre itself—isn't that the truth? We're told that the chiefs were insulted, but personally, what I hear, in the noisy ovation they are given, is the reawakening (unconscious, no doubt) of the latent Boulangism [a right-wing, anti-Germany political movement that advocated militarism] you are still infected with. The blood that flows in your veins is not yet republican blood. Whenever any plumes and ribbons go parading by, your heart beats faster! Whenever a king comes along, you fall in love with him! You're not thinking of your army at all—you just want to go to bed with its general! How very remote from all that the Dreyfus Affair is! While General Billot was being cheered in the Chamber, I could see the shadow of the sabre taking shape on the wall. France, if you're not careful, you're heading straight for dictatorship.

And in addition, this Dreyfus Affair reveals in broad daylight all the nasty politicking that goes on in smoke-filled back rooms; it besmirches the parliamentary system and will do it in. Unfortunately for the Affair, it comes towards the end of the legislative session, when there are only three or four months left in which to rig the next session. Naturally, the government that is currently in power wants to hold elections, and the Deputies are just as determined to get themselves re-elected. So, rather than let portfolios out of their grasp, rather than compromise their chances of election, they are determined to take the most extreme steps. A drowning man could not cling more convulsively to his lifebuoy. That is the crux, that's what explains everything. First of all, the government's extraordinary attitude in this Dreyfus Affair, its silence, its hesitation, the wicked deed it is doing by letting the country agonize amid imposture, when it was the government itself that was responsible for establishing the truth; and then the limp and cowardly lack of interest shown by the Deputies, who pretend they don't know what's going on. They are afraid of only one thing: compromising their re-election by alienating the people, whom they believe to be anti-Semitic. They keep telling you over and over again, 'Ah, if only you would hold elections! you'd see how the government and the parliament would settle the Dreyfus Affair in twenty-four hours!' And that's how the smoke-filled-room politicking of the parliamentary system brings low a great people! . . .

The Evil of Anti-Semitism

France, I beg of you, come to your senses, be yourself again, be that great country, France.

Singlehanded, anti-Semitism has perpetrated two sensational affairs that have done so much harm to France: the Panama Canal

Affair and the Dreyfus Affair. Do you remember how the foul press used denunciations, spread abominable gossip and published false or stolen documents until the Panama business became a hideous ulcer that ate away at the country for years, making it weaker and weaker? The press drove public opinion wild; the entire nation was perverted, intoxicated; the entire nation saw red, demanded figures, decreed that the Parliament was rotten to the core and clamoured for a mass execution of all its members. Ah, if only Arton would come back! if only Arton would speak up! Well, come back he did, and speak up he did. Then all the lies spread by the foul press were revealed as hollow fabrications. And in fact public opinion abruptly swung to the opposite extreme, refusing so much as to suspect a single guilty person and insisting on a mass acquittal! Now, I don't suppose that everyone actually had a clear conscience, for what had happened was what happens in every Parliament in the world when big companies are handling millions and millions. But ultimately the public was nauseated by all the filth; too many people had been smeared, too many had been denounced, and finally the public felt an overriding need to wash it all away and breathe pure air. It needed to believe that everyone was innocent.

Well, I predict that this is what will happen with the Dreyfus Affair, the other crime against society perpetrated by anti-Semitism. Once again, the foul press is saturating the public too heavily with lies and calumny. The press is over determined to turn decent people into knaves, and knaves into decent people. It is spreading so many idiotic stories that finally even children stop believing them. It is arousing too many denials, going too much against the grain of sheer common sense and integrity. One of these days the public will suddenly gag on all the filth it has been fed. It is bound to happen. And just as in the Panama Canal scandal, you'll see that in this Dreyfus Affair as well, the public will bring its weight to bear. In an outpouring of sovereign generosity, the public will decide there are to be no more traitors; it will call for truth and justice. Thus, anti-Semitism will be tried and sentenced for its evil deeds, for the two mortally dangerous follies it has led this country into and for the loss of dignity and health this country has suffered as a result.

France, that is why I beseech you, come back to your senses now; do not wait any longer. The truth cannot be told to you now, since the matter is before the courts and we have no choice but to believe they will reveal the truth. At this point only the judges have the floor. No one else will have a duty to speak out unless the judges fail to reveal the whole truth, the simple truth. But don't you already suspect what that truth is? First there was a mistake; then came all the misdeeds for the sake of hiding that mistake. The facts have been so eloquent that every phase of the investigation has been an admission: Major Esterhazy given the benefit of inexplicable protection again and again; Colonel Picquart treated as if he were the guilty party and suffering outrageous insults, the Ministers playing on words, the unofficial newspapers lying in their teeth, the initial inquiry merely groping about with desperate slowness. Don't you agree there's a rotten stench? Doesn't it reek of dead bodies? Don't you agree they must really have a great many things to hide since they openly let themselves be defended by all the scoundrels in Paris, while decent people are clamouring for a ray of light, even though their doing so prevents them from leading peaceful lives?

France Must Wake Up

France, awaken! Think of your glory! Is it possible that your liberal middle class and your emancipated people do not realize what a senseless scandal they have been tricked into? I cannot believe they are accomplices to it. They must have been fooled, since they are not aware that two things lie behind it: military dictatorship and reactionary clericalism. France, is that what you want? Do you want to jeopardize all that you have paid for so dearly: religious tolerance, equality of justice for all, fraternal solidarity among all of your citizens? If there is the slightest doubt about Dreyfus's guilt and if in spite of that doubt you leave him to languish as an outcast, that is enough to compromise your glorious conquest of law and liberty forever. Will there really be only a handful of us to say these things out loud? Won't we be joined by all of your children, all the decent individuals, all the free spirits, all the generous souls who founded the Republic and should be trembling to see it in such frightful danger?

France, those are the people I appeal to! They must group together! They must write; they must speak up. They must work with us to enlighten the little people, the humble people who are being poisoned and forced into delirium. The nation's soul and energy cannot be triumphant except in a climate of fairness and generosity.

I worry about one thing only: that light may not be shed in full, and may not be shed immediately. A judgement behind closed doors, following on a secret investigation, would not put an end to anything. Only then would the Affair truly begin: for people would have to speak up, since keeping silent would make them accomplices. What folly it is, to think that one can prevent History from being written! Well, the history of this affair *shall* be written, and not one person with any responsibility in it, no matter how slight, will go unpunished.

And ultimately, France, this will enhance your glory. At heart, I do not fear for you: they will try to undermine your sanity and your health, but in vain. You are the future. You will always reawaken, you will always triumph amid truth and justice!

**Wilhelm Röntgen Discovers X-rays:
November 8, 1895**

Röntgen Discovery Changes Science

by H.J.W. Dam

Wilhelm Röntgen's discovery of the X-ray on November 8, 1895, had a rippling effect almost immediately. Soon after the publication of Röntgen's research, scientists engaged in further study and a captivated public wished to know more. People were fascinated—and sometimes fearful—of seeing their own skeletons. The discovery was perceived as strange yet innovative, even before the practical uses of X-rays were fully understood.

In response to the public interest in the scientist's discovery, *McClure's* magazine published an interview by H.J.W. Dam with Röntgen in April 1896. Dam's story includes background information on Röntgen's career as well as the details surrounding the initial discovery. Dam acknowledges that the discovery of X-rays, unlike much of physics research, is accessible to the general population. Scientists and nonscientists alike are able to imagine the potential uses for the X-ray.

I n all the history of scientific discovery there has never been, perhaps, so general, rapid, and dramatic an effect wrought on the scientific centers of Europe as has followed . . . upon an announcement made to the Würzburg Physio-Medical Society,

H.J.W. Dam, "The New Marvel in Photography," *McClure's*, April 1896.

at their December meeting, by Professor William Konrad Röntgen, professor of physics at the Royal University of Würzburg.
The first news which reached London was by telegraph from Vienna to the effect that a Professor Röntgen, until then the pos-
sessor of only a local fame in the town mentioned, had discovered a new kind of light, which penetrated and photographed
through everything. This news was received with a mild interest,
some amusement, and much incredulity; and a week passed.
Then, by mail and telegraph, came daily clear indications of the
stir which the discovery was making in all the great line of universities between Vienna and Berlin. Then Röntgen's own report
arrived, so cool, so business-like, and so truly scientific in character, that it left no doubt either of the truth or of the great importance of the preceding reports. To-day, four weeks after the
announcement, Röntgen's name is apparently in every scientific
publication issued of this week in Europe; and accounts of his
experiments, of the experiments of others following his method,
and of theories as to the strange new force which he has been the
first to observe, fill pages of every scientific journal that comes
to hand. And before the necessary time elapses for this article to
attain publication in America, it is in all ways probable that the
laboratories and lecture-rooms of the United States will also be
giving full evidence of this contagious arousal of interest over a
discovery so strange that its importance cannot yet be measured,
its utility be even prophesied, or its ultimate beliefs be even
vaguely foretold.

The Rays Explained

The Röntgen rays are certain invisible rays resembling, in many
respects, rays of light, which are set free when a high pressure
electric current is discharged through a vacuum tube. A vacuum
tube is a glass tube from which all the air, down to one-millionth
of an atmosphere, has been exhausted after the insertion of a platinum wire in either end of the tube for connection with the two
poles of a battery or induction coil. When the discharge is sent
through the tube, there proceeds from the anode—that is, the
wire which is connected with the positive pole of the battery—
certain bands of light, varying in color with the color of the glass.
But these are insignificant in comparison with the brilliant glow
which shoots from the cathode, or negative wire. This glow ex-
cites brilliant phosphorescence in glass and many substances,

and these "cathode rays," as they are called, were observed and studied by [German physicist Heinrich] Hertz, and more deeply by his assistant, Professor [Philipp] Lenard. Lenard having, in 1894, reported that the cathode rays would penetrate thin films of aluminum, wood, and other substances, and produce photographic results beyond. It was left, however, for Professor Röntgen to discover that during the discharge another kind of rays are set free, which differ greatly from those described by Lenard as cathode rays. The most marked difference between the two is the fact that Röntgen rays are not deflected by a magnet, indicating a very essential difference, while their range and penetrative power are incomparably greater. In fact, all those qualities which have lent a sensational character to the discovery of Röntgen's rays were mainly absent from those of Lenard, to the end that, although Röntgen has not been working in an entirely new field, he has by common accord been freely granted all the honors of a great discovery.

Exactly what kind of a force Professor Röntgen has discovered he does not know. As will be seen below, he declines to call it a new kind of light, or even a new form of electricity. He has given it the name of the X rays. Others speak of it as the Röntgen rays. Thus far its results only, and not its essence, are known. In the terminology of science it is generally called "a new mode of motion," or, in other words, a new force. As to whether it is or not actually a force new to science, or one of the known forces masquerading under strange conditions, weighty authorities are already arguing. More than one eminent scientist has already affected to see in it a key to the great mystery of the law of gravity. All who have expressed themselves in print have admitted, with more or less frankness, that, in view of Röntgen's discovery, science must forthwith revise, possibly to a revolutionary degree, the long accepted theories concerning the phenomena of light and sound. That the X rays, in their mode of action, combine a strange resemblance to both sound and light vibrations, and are destined to materially affect, if they do not greatly alter, our views of both phenomena, is already certain; and beyond this is the opening into a new and unknown field of physical knowledge, concerning which speculation is already eager, and experimental investigation already in hand, in London, Paris, Berlin, and, perhaps, to a greater or less extent, in every well-equipped physical laboratory in Europe.

Popular Understanding and Use

This is the present scientific aspect of the discovery. But, unlike most epoch-making results from laboratories, this discovery is one which, to a very unusual degree, is within the grasp of the popular and non-technical imagination. Among the other kinds of matter which these rays penetrate with ease is the human flesh. That a new photography has suddenly arisen which can photograph the bones, and, before long, the organs of the human body; that a light has been found which can penetrate, so as to make a photographic record, through everything from a purse or a pocket to the walls of a room or a house, is news which cannot fail to startle everybody. That the eye of the physician or surgeon, long baffled by the skin, and vainly seeking to penetrate the unfortunate darkness of the human body, is now to be supplemented by a camera, making all the parts of the human body as visible, in a way, as the exterior, appears certainly to be a greater blessing to humanity than even the Listerian antiseptic system of surgery; and its benefits must inevitably be greater than those conferred by [English surgeon and scientist Joseph] Lister, great as the latter have been. Already, in the few weeks since Röntgen's announcement, the results of surgical operations under the new system are growing voluminous. In Berlin, not only new bone fractures are being immediately photographed, but joined fractures, as well, in order to examine the results of recent surgical work. In Vienna, imbedded bullets are being photographed, instead of being probed for, and extracted with comparable ease. In London, a wounded sailor, completely paralyzed, whose injury was a mystery, has been saved by the photographing of an object imbedded in the spine, which, upon extraction, proved to be a small knife-blade. Operations for malformations, hitherto obscure, but now clearly revealed by the new photography, are already becoming common, and are being reported from all directions. Professor Czermak of Graz has photographed the living skull, denuded of flesh and hair, and has begun the adaptation of the new photography to brain study. The relation of the new rays to thought rays is being eagerly discussed in what may be called the non-exact circles and journals; and all that numerous group of inquirers into the occult, the believers in clairvoyance, spiritualism, telepathy, and kindred orders of alleged phenomena, are confident of finding in the new force long-sought facts in proof of their claims. Professor Neusser in Vienna has photographed gall-stones in the liver of one patient (the stone showing snow-white in the negative), and a stone in the bladder of another patient. His results so far induce him to announce that all the organs of the human body can, and will, shortly, be photographed. [French physician Odilion] Lannelongue of Paris has exhibited to the Academy of Science photographs of bones showing inherited tuberculosis which had not otherwise revealed itself. Berlin has already a society of forty for the immediate prosecution or researches into both the character of the new force and its physiological possibilities. In the next few weeks these strange announcements will be trebled or quadrupled, giving the best evidence from all quarters of the great future that awaits the Röntgen rays, and the startling impetus to the universal search for knowledge that has come at the close of the nineteenth century from the modest little laboratory in the Pleicher Ring at Würzburg.

Meeting the Discoverer

On instruction by cable from the editor of this magazine, on the first announcement of the discovery, I set out for Würzburg to see the discoverer and his laboratory. I found a neat and thriving Bavarian city of forty-five thousand inhabitants, which, for some ten centuries, has made no salient claim upon the admiration of the world, except for the elaborateness of its mediaeval castle and the excellence of its local beer. Its streets were adorned with large numbers of students, all wearing either scarlet, green, or blue caps, and an extremely serious expression, suggesting much intensity either in the contemplation of Röntgen rays or of the beer aforesaid. All knew the residence of Professor Röntgen (pronunciation: "Renken"), and directed me to the "Pleicher Ring." The various buildings of the university are scattered in different parts of Würzburg, the majority being in the Pleicher Ring, which is a fine avenue, with a park along one side of it, in the center of the town. The Physical Institute, Professor Röntgen's particular domain, is a modest building of two stories and basement, the upper story constituting his private residence, and the remainder of the building being given over to lecture rooms, laboratories, and their attendant offices. At the door I was met by an old serving-man of the idolatrous order, whose pain was apparent when I asked for "Professor" Röntgen, and he gently corrected me with "Herr Doctor Röntgen." As it was evident, however, that we referred to the same person, he conducted me along a wide, bare hall, running the length of the building, with blackboards and charts on the walls. At the end he showed me into a small room on the right. This contained a large table desk, and a small table by the window, covered with photographs, while the walls held rows of shelves laden with laboratory and other records. An open door led into a somewhat larger room, perhaps twenty feet by fifteen, and I found myself gazing into a laboratory which was the scene of the discovery—a laboratory which, though in all ways modest, is destined to be enduringly historical.

There was a wide table shelf running along the farther side, in front of the two windows, which were high, and gave plenty of light. In the center was a stove; on the left, a small cabinet, whose shelves held the small objects which the professor had been using. There was a table in the left-hand corner; and another small table—the one on which living bones were first photographed—was near the stove, and a Rhumkorff coil was on the right. The lesson of the laboratory was eloquent. Compared, for instance, with the elaborate, expensive, and complete apparatus of, say, the University of London, or of any of the great American universities, it was bare and unassuming to a degree. It mutely said that in the great march of science it is the genius of man, and not the perfection of appliances, that breaks new ground in the great territory of the unknown. It also caused one to wonder at and endeavor to imagine the great things which are to be done through elaborate appliances with the Röntgen rays—a field in which the United States, with its foremost genius in invention, will very possibly, if not probably, take the lead—when the discoverer himself had done so much with so little. Already, in a few weeks, a skilled London operator, Mr. A.A.C. Swinton, has reduced the necessary time of exposure for Röntgen photographs from fifteen minutes to four. He used, however, a Tesla oil coil, discharged by twelve half-gallon Leyden jars, with an alternating current of twenty thousand volts pressure. Here were no oil coils, Leyden jars, or specially elaborate and expensive machines. There were only a Rhumkorff coil and Crookes (vacuum) tube and the man himself.

Professor Röntgen entered hurriedly, something like an amiable gust of wind. He is a tall, slender, and loose-limbed man, whose whole appearance bespeaks enthusiasm and energy. He wore a dark blue sack suit, and his long, dark hair stood straight up from his forehead, as if he were permanently electrified by his own enthusiasm. His voice is full and deep, he speaks rapidly, and, altogether, he seems clearly a man who, once upon the track of a mystery which appeals to him, would pursue it with unremitting vigor. His eyes are kind, quick, and penetrating; and there is no doubt that he much prefers gazing at a Crookes tube to beholding a visitor, visitors at present robbing him of much valued time. The meeting was by appointment, however, and his greeting was cordial and hearty. In addition to his own language he speaks French well and English scientifically, which is different from speaking it popularly. These three tongues being more or less within the equipment of this visitor, the conversation proceeded on an international or polyglot basis, so to speak, varying at necessity's demand.

"Seeing" the Invisible Rays

It transpired, in the course of inquiry, that the professor is a married man and fifty years of age, though his eyes have the enthusiasm of twenty-five. . . . He has been at Würzburg about seven years, and had made no discoveries which he considered of great importance prior to the one under consideration. These details were given under good-natured protest, he failing to understand why his personality should interest the public. He declined to admire himself or his results in any degree, and laughed at the idea of being famous. The professor is too deeply interested in science to waste any time in thinking about himself. His emperor had feted, flattered, and decorated him, and he was loyally grateful. It was evident, however, that fame and applause had small attractions for him, compared to the mysteries still hidden in the vacuum tubes of the other room.

"Now, then," said he, smiling, and with some impatience, when the preliminary questions at which he chafed were over, "you have come to see the invisible rays?"

"Is the invisible visible?"

"Not to the eye; but its results are. Come in here."

He led the way to the other square room mentioned, and indicated the induction coil with which his researches were made, an ordinary Rhumkorff coil, with a spark of from four to six inches, charged by a current of twenty amperes. Two wires led from the coil, through an open door, into a smaller room on the right. In this room was a small table carrying a Crookes tube connected with the coil. The most striking object in the room, however, was a huge and mysterious tin box about seven feet high and four feet square. It stood on end, like a huge packing-case, its side being perhaps five inches from the Crookes tube.

The professor explained the mystery of the tin box, to the effect that it was a device of his own for obtaining a portable darkroom. When he began his investigations he used the whole room, as was shown by the heavy blinds and curtains so arranged as to exclude the entrance of all interfering light from the windows. In the side of the tin box, at the point immediately against the tube, was a circular sheet of aluminum one millimeter in thickness, and perhaps eighteen inches in diameter, soldered to the surrounding tin. To study his rays the professor had only to turn on the current, enter the box, close the door, and in perfect darkness inspect only such light or light effects as he had a right to consider his own, hiding his light, in fact, not under the Biblical bushel, but in a more commodious box.

"Step inside," he said, opening the door, which was on the side of the box farthest from the tube. I immediately did so, not altogether certain whether my skeleton was to be photographed for general inspection, or my secret thoughts held up to light on a glass plate.

"You will find a sheet of barium paper on the shelf," he added, and then went away to the coil. The door was closed, and the interior of the box became black darkness. The first thing I found was a wooden stool, on which I resolved to sit. Then I found the shelf on the side next to the tube, and then the sheet of paper prepared with barium platinocyanide. I was thus being shown the first phenomenon which attracted the discoverer's attention and led to the discovery, namely, the passage of rays, themselves wholly invisible, whose presence was only indicated by the effect they produced on a piece of sensitized photographic paper.

A moment later, the black darkness was penetrated by the rapid snapping sound of the high-pressure current in action, and I knew that the tube outside was glowing. I held the sheet vertically on the shelf, perhaps four inches from the plate. There was no change, however, and nothing was visible.

"Do you see anything?" he called.

"No."

"The tension is not high enough;" and he proceeded to increase the pressure by operating an apparatus of mercury in long vertical tubes acted upon automatically by a weight lever which stood near the coil. In a few minutes the sound of the discharge again began, and then I made my first acquaintance with the Röntgen rays.

The moment the current passed, the paper began to glow. A yellowish-green light spread all over its surface in clouds, waves, and flashes. The yellow-green luminescence, all the stranger and stronger in the darkness, trembled, wavered, and floated over the paper, in rhythm with the snapping of the discharge. Through the metal plate, the paper, myself, and the tin box, the invisible rays were flying, with an effect strange, interesting, and uncanny. The metal plate seemed to offer no appreciable resistance to the flying force, and the light was as rich and full as if nothing lay between the paper and the tube.

"Put the book up," said the professor.

I felt upon the shelf, in the darkness, a heavy book, two inches in thickness, and placed this against the plate. It made no difference. The rays flew through the metal and the book as if neither had been there, and the waves of light, rolling cloud-like over the paper, showed no change in brightness. It was a clear, material illustration of the ease with which paper and wood are penetrated. And then I laid book and paper down, and put my eyes against the rays. All was blackness, and I neither saw nor felt anything. The discharge was in full force, and the rays were flying through my head, and for all I knew, through the side of the box behind me. But they were invisible and impalpable. They gave no sensation whatever. Whatever the mysterious rays may be, they are not to be seen, and are to be judged only by their works.

Inside the Human Body

I was loath to leave this historical tin box, but time pressed. I thanked the professor, who was happy in the reality of his discovery and the music of his sparks. Then I said: "Where did you first photograph living bones?"

"Here," he said, leading the way into the room where the coil stood. He pointed to a table on which stood another—the latter a small short-legged wooden one with more the shape and size of a wooden seat. It was two feet square and painted coal black. I viewed it with interest. I would have bought it, for the little table on which light was first sent through the human body will some day be a great historical curiosity; but it was *"nicht zu verkaufen"* ["not for sale"]. A photograph of it would have been a consolation, but for several reasons one was not to be had at present. However, the historical table was there, and was duly inspected. "How did you take the first hand photograph?" I asked.

The professor went over to a shelf by the window, where lay a number of prepared glass plates, closely wrapped in black paper. He put a Crookes tube underneath the table, a few inches from the under side of its top. Then he laid his hand flat on the top of the table, and placed the glass plate loosely on his hand.

"You ought to have your portrait painted in that attitude," I suggested.

"No, that is nonsense," said he, smiling.

"Or to be photographed." This suggestion was made with a deeply hidden purpose.

The rays from the Röntgen eyes instantly penetrated the deeply hidden purpose.

"Oh, no," said he; "I can't let you make pictures of me. I am too busy." Clearly the professor was entirely too modest to gratify the wishes of the curious world.

"Now, Professor," said I, "will you tell me the history of the discovery?"

"There is no history," he said. "I have been for a long time interested in the problem of the cathode rays from a vacuum tube as studied by Hertz and Lenard. I had followed theirs and other researches with great interest, and determined, as soon as I had the time, to make some researches of my own. This time I found at the close of last October. I had been at work for some days when I discovered something new."

"What was the date?"

"The eighth of November."

"And what was the discovery?"

"I was working with a Crookes tube covered by a shield of black cardboard. A piece of barium platinocyanide paper lay on the bench there. I had been passing a current through the tube, and I noticed a peculiar black line across the paper."

"What of that?"

"The effect was one which could only be produced, in ordinary parlance, by the passage of light. No light could come from the tube, because the shield which covered it was impervious to any light known, even that of an electric arc."

"And what did you think?"

"I did not think; I investigated. I assumed that the effect must have come from the tube, since its character indicated that it could come from nowhere else. I tested it. In a few minutes there was no doubt about it. Rays were coming from the tube which had a luminescent effect upon the paper. I tried it successfully at greater and greater distances, even at two meters. It seemed at first a new kind of invisible light. It was clearly something new, something unrecorded.

"Is it light?"

"No."

"Is it electricity?"

"Not in any known form."

"What is it?"

"I don't know."

And the discoverer of the X rays thus stated as calmly his ignorance of their essence as has everybody else who has written on the phenomena thus far.

"Having discovered the existence of a new kind of rays, I of course began to investigate what they would do."

He took up a series of cabinet-sized photographs. "It soon appeared from tests that the rays had penetrative power to a degree hitherto unknown. They penetrate paper, wood, and cloth with ease; and the thickness of the substance made no perceptible difference, within reasonable limits."

He showed photographs of a box of laboratory weights of platinum, aluminum, and brass, they and the brass hinges all having been photographed from a closed box, without any indication of the box. Also a photograph of a coil of fine wire, wound on a wooden spool, the wire having been photographed, and the wood omitted.

"The rays," he continued, "passed through all the metals tested, with a facility varying, roughly speaking, with the density of the metal. These phenomena I have discussed carefully in my report to the Würzburg society, and you will find all the technical results therein stated."

He showed a photograph of a small sheet of zinc. This was composed of smaller plates soldered laterally with solders of different proportions. The differing lines of shadow, caused by the difference in solders, were visible evidence that a new means of detecting flaws and chemical variations in metals had been found. A photograph of a compass showed the needle and dial taken through the closed brass cover. The markings of the dial were in red metallic paint, and thus interfered with the rays, and were reproduced.

"Since the rays had this great penetrative power, it seemed natural that they should penetrate flesh, and so it proved in photographing the hand I showed you."

What the Rays Mean for the Future

A detailed discussion of the characteristics of his rays the professor considered unprofitable and unnecessary. He believes, though, that these mysterious radiations are not light, because their behavior is essentially different from that of light rays, even those light rays which are themselves invisible. The Röntgen rays cannot be reflected by reflecting surfaces, concentrated by lenses, or refracted or diffracted. They produce photographic action on a sensitive film, but their action is weak as yet, and herein lies the first important field of their development. The professor's exposures were comparatively long—an average of fifteen minutes in easily penetrable media, and half an hour or more in photographing the bones of the hand. Concerning vacuum tubes, he said that he preferred the Hittorf, because it had the most perfect vacuum, the highest degree of air exhaustion being the consummation most desirable. In an answer to a question, "What of the future?" he said:

"I am not a prophet, and I am opposed to prophesying. I am pursuing my investigations, and as fast as my results are verified I shall make them public."

"Do you think the rays can be so modified as to photograph the organs of the human body?"

In answer he took up the photograph of the box of weights. "Here are already modifications," he said, indicating the various degrees of shadow produced by the aluminum, platinum, and brass weights, the brass hinges, and even the metallic stamped lettering on the cover of the box, which was faintly perceptible.

"But Professor Neusser has already announced that the photographing of the various organs is possible."

"We shall see what we shall see," he said. "We have the start now; the developments will follow in time."

"You know the apparatus for introducing the electric light into the stomach?"

"Yes."

"Do you think that this electric light will become a vacuum

tube for photographing, from the stomach, any part of the abdomen or thorax?"

The idea of swallowing a Crookes tube, and sending a high frequency current down into one's stomach, seemed to him exceedingly funny.

"When I have done it, I will tell you," he said, smiling, resolute in abiding by results.

"There is much to do, and I am busy, very busy," he said in conclusion. He extended his hand in farewell, his eyes already wandering toward his work in the inside room. And his visitor promptly left him; the words, "I am busy," said in all sincerity, seemed to describe in a single phrase the essence of his character and the watchword of a very unusual man.

Returning by way of Berlin, I called upon Herr Spiess of the Urania, whose photographs after the Röntgen method were the first made public, and have been the best seen thus far. The Urania is a peculiar institution, and one which it seems might be profitably duplicated in other countries. It is a scientific theater. By means of the lantern and an admirable equipment of scientific appliances, all new discoveries, as well as ordinary interesting and picturesque phenomena, when new discoveries are lacking, are described and illustrated daily to the public, who pay for seats as in an ordinary theater, and keep the Urania profitably filled all the year round. Professor Spiess is a young man of great mental alertness and mechanical resource. It is the photograph of a hand, his wife's hand, which illustrates, perhaps better than any illustration in this article, the clear delineation of the bones which can be obtained by the Röntgen rays. In speaking of the discovery he said:

"I applied it, as soon as the penetration of flesh was apparent, to the photograph of a man's hand. Something in it had pained him for years, and the photograph at once exhibited a small foreign object, as you can see"; and he exhibited a copy of the photograph in question.

"The speck there is a small piece of glass, which was immediately extracted, and which, in all probability, would have otherwise remained in the man's hand to the end of his days."

All of which indicates that the needle which has pursued its travels in so many persons, through so many years, will be suppressed by the camera.

"My next object is to photograph the bones of the entire leg," continued Herr Spiess. "I anticipate no difficulty, though it requires some thought in manipulation."

It will be seen that the Röntgen rays and their marvelous practical possibilities are still in their infancy. The first successful modification of the action of the rays so that the varying densities of bodily organs will enable them to be photographed, will bring all such morbid growth as tumors and cancers into the photographic field, to say nothing of vital organs which may be abnormally developed or degenerate. How much this means to medical and surgical practice it requires little imagination to conceive. Diagnosis, long a painfully uncertain science, has received an unexpected and wonderful assistant; and how greatly the world will benefit thereby, how much pain will be saved, and how many lives saved, the future can only determine. In science a new door has been opened where none was known to exist, and a side-light on phenomena has appeared, of which the results may prove as penetrating and astonishing as the Röntgen rays themselves. The most agreeable feature of the discovery is the opportunity it gives for other hands to help; and the work of these hands will add many new words to the dictionaries, many new facts to science, and, in the years long ahead of us, fill many more volumes than there are paragraphs in this brief and imperfect account.

10 A Motion Picture Is Shown to a Public Audience: December 28, 1895

The Lumière Brothers Present a Moving-Picture Show

by Eric Rhode

Photography and moving pictures underwent a series of improvements in the late nineteenth century. Many inventors elaborated on the work and ideas of other inventors. For example, in 1893 Thomas Edison invented a device that allowed one person at a time to view a moving image. The Lumière brothers, Auguste and Louis, enhanced this discovery by projecting moving pictures in a way that would allow many people to view them at once.

With the notion of a larger audience in mind, Louis Lumière developed the cinematograph, which continuously moved sections of film across a lens at a rate of twelve frames per second. Additionally, the device could produce both positive and negative images on film. The camera was used for the first time in 1894, when workers leaving the Lumière factory were filmed. Soon the Lumière brothers decided to show this and other short films to an audience. On December 28, 1895, at the Grand Café in Paris, Auguste and Louis Lumière held the first public showing of motion pictures.

The following selection by Eric Rhode highlights the role of the Lumière brothers in the history of film. According to Rhode, the achievement of inventing the cinematograph cannot be overrated.

Film historian and broadcaster Eric Rhode of Great Britain wrote a work surveying the history of film.

T he most decisive event in these early years, the first public showing of the *cinématographe*, was only accidentally related to the idea of film as an entertainment industry. Antoine Lumière, a successful photographer, and his eldest son, Louis, had invented a type of photographic plate that was greatly in demand. The family had prospered to such an extent that their factory at Lyons was to employ a staff of over three hundred people: it was second only to George Eastman's at Rochester N.Y. Recognizing the commercial value of the kinetoscope, Antoine was dismayed by the prohibitive charges [Thomas] Edison [who had invented a motion-picture camera] was making on the sale of film and asked Louis to invent something like it. Louis, in the event, came up with something better: a piece of equipment which was both camera and projector and had a novel clawdrive for moving the film strip.

Inventing the Cinema

Louis Lumière was modest about his achievement. 'What did I do? It was in the air. The work of [Zacharias] Janssen [inventor of the compound microscope in 1590], of [Étienne-Jules] Marey [inventor of several motion-picture devices in the 1880s] and his followers had one day or another to lead to its discovery.' (A generous tribute: one of Marey's assistants, Georges Demenÿ, unjustly sued the Lumières for having filched the idea of the *cinématographe* from his phonoscope.) At the same time Louis had no doubt about the importance of his invention: 'Other machines may have preceded mine. But they didn't work.' Georges Sadoul has argued that it was Louis's care for detail which made him the true originator of the cinema. His camera (built by Charles Moisson) was a highly accomplished piece of engineering, and when fifty years later Georges Sadoul set it in motion once more he found that it functioned perfectly. Its stock of film, so supple and clear, might have just arrived from the laboratory. 'My brother,' said Auguste Lumière, 'invented the cinema in one night'—a night in which Louis was suffering from disturbing dreams and a migraine.

Throughout 1895 the Lumières gave semi-private viewings; and then, at the end of the year, on 28 December, they opened a public auditorium in Paris. Their location was a basement to the Grand Café on the Boulevard des Capucines, the exotically decorated Salon Indien, reached by an awkward spiral staircase. Seats were priced at one franc. During the first few days they aroused little attention; then they caught the public imagination and within weeks were a world wide success. They had shrewdly anticipated this response and over the previous year built up their stock of machines. They had also trained a brigade of cameramen, capable of shooting films as well as projecting them, and sent off these men to foreign countries, where they had soon made a total of twelve hundred single-shot films on many subjects, including the Diamond Jubilee procession in London [celebrating British queen Victoria's reign].

The Lumières recognized their novelty would be short-lived and went in for the quick kill. Their business acumen paid off. On 17 February they opened in London. April saw viewings in Vienna and Geneva: June, in Madrid, Belgrade and New York (where a team of twenty-one operators proved too small for the demand); July, in St Petersburg and Bucharest. During the latter part of the year their cameramen reached Egypt, India, Japan, Australia and many other places. Royalty, and celebrities of many kinds, flocked to their shows. Raff and Gammon, the company distributing Edison's kinetoscope, faced heavy losses.

An Important Development

The importance of this development in the history of communications cannot be overrated. Transcending differences of language and national custom, Lumière's *cinématographe* suddenly made the world a smaller and more ordinary place: the invention of faster forms of transport and of the radio—[Guglielmo] Marconi invented wireless telegraphy in 1895 and the Wright brothers first took flight in 1903—were never to have the same shocking, elating effect. In *The Moment of Cubism*, John Berger has written of how the new technology of the 1890s changed the consciousness of poets like Apollinaire and painters like Picasso and Braque. For a while at least it appeared as though mankind had become a Prometheus who could cheat the gods with impunity. Then came the First World War and put an end to that dream. But even in 1919 a poet like Blaise Cendrars could write about the cinema with an excitement that approached delirium—see, for example, his ciné-poem 'La Fin du monde'. Besides which, in shrinking the world, the *cinématographe* encouraged fashions and trends to spread with lightning speed, obliterating the idiosyncrasies of folk art. In 1909 the Bronco Billy Westerns began to enjoy a vogue in the United States; within months countries as far apart as Italy, Germany and Japan were manufacturing imitation Westerns by the hundred. The coming of the movies helped to undermine the notion of uniqueness, the magical fetishistic quality usually ascribed to visual works of art.

The Public and the Cinema

By 1896 Louis Lumière had begun to lose interest in his camera and was turning his mind to fresh problems. Over the next three decades he was to experiment with the idea of a giant screen— intended for the Paris International Exposition of 1900, but never realized—with wide-gauged film, with the Photorama (a screen encircling the audience), with stereoscopic effects. He also made a serious contribution to colour-plate photography. But he was unattracted to the cinema as an aspect of the entertainment business: and though the Lumière firm did produce fictions and subjects of a biblical nature, Louis had nothing to do with them.

In May 1897, a cinema fire at the Bazar de la Charité, at that time one of the most fashionable of annual events, frightened the public and brought the *cinématographe* into disrepute. An ether lamp, which provided light for the projector, went out of control, and within minutes a provisional auditorium of canvas and wood went up in flames, killing over a hundred people, including some leading society figures and a large number of children. It could be argued, very speculatively, that the sensational nature of this tragedy—it included the death of many aristocrats—was one of the reasons why the middle classes were to stay away from the cinema for over two decades and why it was to be known at first as a working-class entertainment. In 1900 the Lumière family sold off their camera rights to Charles Pathé. Louis Lumière died in 1948 at the age of eighty-four.

How do we account for the astounding popularity of the Lumière invention in 1896? The public had long been aware of moving photographs, of large screens and, in the theatre, of naturalistic detail taken to an absurd extreme. It took pleasure, certainly, in seeing movement of so fluid a kind and in recognizing certain shapes as, say, trembling leaves; but the main reason for its surprise, it could be argued, lay elsewhere.

In one of his films, *The Arrival of a Train*, Louis Lumière had seemingly contrived to break the screen surface so that a train appeared to rush out into the auditorium. This shock depended on catching the audience unawares and on trangressing a propriety in the arts: the distancing of the spectator from the object of his attention. Art, it had been assumed, could be admired if kept at arm's length—the gentleman separated from the players. But this convention, and (related to it) the problem of aggression in the arts, was already being called into question. The Impressionist painters had drawn the spectator's attention to the picture plane (Henri Langlois thinks that Louis had a sense of composition similar to that of Auguste Renoir); and men of the theatre, like [Swedish playwright August] Strindberg and Antoine, had become sensitive to the convention of the 'fourth wall'. which in 1896 the shrieks of Père Ubu, in Alfred Jarry's avant-garde play, were to bring tumbling down. Louis could easily have kept his images within the proscenium convention of the stage, but apparently did not wish to do so. He must have been conscious long before the Paris opening (which he did not attend) that the public would feel the arrival of the train at La Ciotat as an assault.

Another reason for its popularity is more perplexing. We take it for granted that the Victorians found all machines alienatory. In fact, as a writer in the *MacMillan Magazine* claimed in September 1871, their response to photography (and later to the cinema) was quite the opposite:

> Anyone who knows what the worth of family affection is among the lower classes, and who has seen the array of little portraits stuck over a labourer's fireplace, still gathering into one the 'Home' that life is always parting—the boy that has 'gone to Canada', the 'girl out at service', the little one with the golden hair that sleeps under the daisies, the old grandfather in the country—will perhaps feel with me that in counteracting the tendencies, social and industrial, which every day are sapping the healthier family affections, the sixpenny photograph is doing more for the poor than all the philanthropists in the world.

This insight has a human centrality that recalls [English novelist Charles] Dickens. We may find it ironic that a class only recently touched by the hope of literacy should have been the first to give itself over to moving images; but the truth is that the cinema was to be a desperately needed consolation and source of knowledge to the poor, the illiterate and to immigrant communities (as in America) unable to speak the native language.

Powers of Cinema

The *MacMillian Magazine* writer implies that photography can diminish distances of time and place: an astonishing prediction of the cinema's greatest power. He suggests, too, that since photography can redress the social and industrial tendencies of the age, which he equates with death, it has the power, in part, to conquer death. A journalist, writing in 1896 in *La Poste de Paris*, in one of the first reviews of a *cinématographe* show, makes this point more sharply: 'Now that we can photograph our loved ones, not only in stillness, but as they move, as they act, as they make familiar gestures, as they speak—death ceases to be absolute.'

At the same time as the camera appeared to reinforce behaviourism, it confirmed the existence of mental states by shadowing forth images that resembled the images of the mind. Yet this contradiction was not new. While some of [American motion-picture pioneer Eadweard] Muybridge's contemporaries had seen his sequence in *Animal Locomotion* as further evidence for equating man with the machine, Muybridge himself had consulted Edison in 1888 on the possibility of bringing together the photograph and the phonograph to 'reproduce simultaneously, in the presence of an audience, visible actions and audible words' so as to entertain the public 'long after the original participants shall have passed away'. And Marey, for all his concern with time-and-motion studies, was to recognize the poignancy of passing time. 'Each of us', he wrote, 'has experienced the disastrous effect of time on memory. Who has not, on returning to a former place, been undisturbed by the false memory he has kept of it?'

This desire to defy death and the erosions of time had its darker side. By 1861, claims Edgar Morin in *Le Cinéma ou l'homme imaginaire*, spiritualists and faith-healers had begun to use the photograph in place of wax figurines as a talisman for their spells and prognostics. And while Méliès, the conjuror, was soon to include film among the more amusing of his tricks, German directors just before the First World War were to reveal in certain screened images an inherently uncanny quality that was to be less innocent. There were other worlds, it had to be remembered, than the one posited by naturalistic doctrine, and not all of them were to be encouraging. Film opened the way to new nostalgias, modern sentimentalities and sophisticated forms of fetishism.

More positively, the fact that a well-known bustling street might be enshrined on celluloid without losing its vitality conferred an objective standing on the images of memory, as D.W. Griffith, more perhaps than any other director, was to realize. What had previously been seen only in the mind's eye—the dead friend, or the place long unvisited—was now no longer a private (and so perhaps delusive) experience. Marcel Proust, most extensive chronicler of consciousness of his time, within a few years was to begin writing his great novel *A la recherche du temps perdu;* and it was no accident that this prolonged essay in memory should have been woven with numerous tropes picked from the field of physiological optics, from stereoscopy and from a study of the magic lantern and the kaleidoscope. In a subterranean fashion, Louis Lumière confirmed the existence of psychic reality: how the projection of images could become an extension of the mind's eye. It was this confirmation, perhaps, that intuitively moved the first film audiences.

The first Lumière film performance has stood up well to the test of time. Formally simple (each consists of a single, well-composed static shot), at least four of them are intricate in content. *The Card Game*, Cézanne-like [in the style of French impressionist Paul Cézanne] has two men playing cards, another man opening a bottle and filling some glasses, and a capering waiter: a multiplicity of action that keeps the eye engaged and, in fact, offers the eye more than it can absorb at one viewing; it obliges the spectator to forget the camera and to become involved in the content of the film. In the 1930s by similar means, more fully developed, Jean Renoir was to create an impression of realism.

Other Examples of Early Films

At least three other shorts made in 1895 challenge the spectator's powers of observation by concentrating on fast-moving crowds: workers leaving the Lumière factory, with not only hundreds of figures, but bicycles and a dog to hold his attention; members of a photographic congress (a fascinating variety of faces and clothes) disembarking at Neuville-sur-Saône, with Janssen in the lead; and the celebrated arrival of a train at La Ciotat station, with again a crowd of people, among them the hauntingly beautiful figure of a woman in a wide-brimmed hat. In discovering unexpected images in everyday experience, these films discovered a source often to be returned to at times of cinematic renewal, as in the Kino-Eye investigations, or neo-realism.

Their shortness (they last about a minute) and their pith give them the sharp memorability of certain dreams. Is it no more than a coincidence that [Austrian neurologist Sigmund] Freud discovered the key to psychoanalysis in the year that Lumière created these films? 'The Interpretation of Dreams', Freud wrote to Wilhelm Fleiss, 'was finished in all essentials at the beginning of 1896.' In psychoanalysis the crucial concept of transference, in which the patient 'projects' his fantasies onto what he feels to be the impersonal screen-like mind of the analyst, finds its analogy in the cinema and other nineteenth-century optical inventions to the same extent that psychoanalysis itself had developed out of experimental psychology.

If film humour originated in the music-hall gag, it did so perhaps because the films were so short: part of the pleasure of these gags lies in the tension of wondering whether the film will run out before the joke has been completed. The first one on record, rudimentary and rather feeble, was made at the studio Edison had built in the garden of his laboratory, a revolving, heliotropic prefabricated building that bore the nickname of 'Black Maria' (it looked like a police patrol wagon). Called *Fred Ott's Sneeze* this gag still delights audiences—in part, probably, because the build-up of the sneeze and its discharge is timed to fit the film's length. On the other hand, Louis's more sophisticated *L'Arroseur arrosé (Watering the Gardener)* continues beyond the gag—that is, of the boy stepping on the hose and the gardener having his face soaked: the spectator sees the gardener spank the boy and then return to his duties. Perhaps Louis did not want us to identify with the boy, as some of his successors ([French filmmaker Ferdinand] Zecca, actor Charlie Chaplin) might have wished. A rough sort of justice, you could say, brings stability back to the image.

Pioneers in the Early Years

The pioneers dwelt under the aegis of science. At the same time, many of the inventors were isolated men, unprotected by any kind of institution, lacking financial backing, often on the verge of despair. Their intuitions seldom had more than a random connection with theory. Edison, an autodidact, was obliged to employ a mathematician, Francis R. Upton, who had studied for a year under [German physicist, anatomist, and physiologist Hermann] Helmholtz. (When Edison met Helmholtz they had little to say to each other.) Edison nicknamed Upton 'Culture'. He despised theory and scorned experiments that had no immediate application to industry.

But through his achievements he had, like the Lumières, built up material resources: a fine workshop at Menlo Park, a large staff, the provisional assistance of Wall Street bankers. Other pioneers lived a hand-to-mouth, lodging-house existence: William Friese-Greene, forever taking out patents, forever leaving his inventions unfinished, jailed for bankruptcy, dying at a film congress with less than two shillings in his pocket; the debonair Major Woodville Latham and his two sons, failed salesmen, failed businessmen, failed showmen—yet claimants to the Latham loop, still used on the modern projector: George Eastman, a multi-millionaire admittedly, who shot himself, leaving a note— 'My work is done. Why wait?'; Emile Reynaud, who allegedly threw his machines into the Seine; Muybridge, impractical and a little mad, jailed briefly for murdering his wife's lover: Louis Le Prince, working on the same lines as the Lumières and perhaps the first to think of perforating the film strip, who mysteriously disappeared—perhaps in flight from his wife—while travelling by train from Dijon to Paris; Jean LeRoy, the original film entrepreneur, who founded a cinematograph company in New York and then vanished into obscurity. 'The cares and anxieties of being an inventor,' wrote Alexander Graham Bell, originator of the telephone, 'seem more than flesh and blood can stand.' The tricksters, the fast talkers, the cranks, the defeated: all of them took part in the gold rush to invent the cinema. Yet for years the cinema was more likely to lose your fortune than to make it. It hovered between life and death in the nether world of the fairground, the second-class music hall, the beer garden, the penny arcades and the church social.

The Curies Did Revolutionary Research

by Nancy Fröman

Polish-born Marie Curie and her French husband Pierre were pioneers in researching radioactivity (a word coined by Marie) at the end of the nineteenth century. They expanded on the research of Antoine Henri Becquerel, who discovered that the element uranium gave off radiation. Marie investigated her hypothesis that radiation was a property found in atoms. She supposed that other elements might also possess radiation.

The Curies then examined other elements for evidence of radioactivity (the property of having radiation) and in the course of their research, they discovered two new radioactive elements. The first element, polonium, was named for Marie's native Poland. The Curies submitted their findings to the Academy of Sciences, and on July 18, 1898, the scientific community learned of their research. At a meeting of the Academy of Sciences on December 26, 1898, a paper was read announcing the Curies' second discovery: radium.

Between 1898 and 1902, the Curies produced, either together or independently, thirty-two scientific papers. One of them revealed that diseased cells exposed to radium were destroyed faster than healthy cells subject to the same exposure. This important revelation had a great impact on medicine and the treatment of cancer.

The following lecture by Nancy Fröman recounts the significant contributions made to the study of physics at the end of the nine-

Nancy Fröman, lecture at the Royal Academy of Sciences, translated by Nancy Marshall-Lundén, Stockholm, Sweden, February 28, 1996. Copyright © 1996 by Nancy Marshall-Lundén. Reproduced by permission.

teenth century. The Curies based the direction of their research on these contributions; soon they made important discoveries of their own. Fröman explains how ideas—which she cites as Marie's greatest contribution to conceptual thinking in physics—turned into scientific experimentation with surprising results. The new elements they discovered had a great deal of radioactivity, especially radium. After laborious analyses of various types of ore, the Curies verified their discoveries and elaborated on their research. Fröman reminds us that Marie and Pierre did not wish to patent their ideas, because they believed that their research should be available to everyone. When production of mass quantities of radium began for medical purposes, the Curies gladly provided information about the process.

In 1903 the Curies and Becquerel were awarded the Nobel Prize in Physics for their work on the radiation phenomena.

A t the end of the 19th century, a number of discoveries were made in physics which paved the way for the breakthrough of modern physics and led to the revolutionary technical development that is continually changing our daily lives.

Important Physics Discoveries

Around 1886, Heinrich Hertz demonstrated experimentally the existence of radio waves. It is said that Hertz only smiled incredulously when anyone predicted that his waves would one day be sent round the earth. Hertz died in 1894 at the early age of 37. In September 1895, Guglielmo Marconi sent the first radio signal over a distance of 1.5 km. In 1901 he spanned the Atlantic. Hertz did not live long enough to experience the far-reaching positive effects of his great discovery, nor of course did he have to see it abused in bad television programs. It is hard to predict the consequences of new discoveries in physics.

On November 8, 1895, Wilhelm Conrad Röntgen at the University of Würzburg, discovered a new kind of radiation which he called X-rays. It could in time be identified as the short-wave, high frequency counterpart of Hertz's waves. The ability of the radiation to pass through opaque material that was impenetrable to ordinary light, naturally created a great sensation. Röntgen himself wrote to a friend that initially, he told no one except his wife about what he was doing. People would say, "Röntgen is

out of his mind." On January 1, 1896, he mailed his first announcement of the discovery to his colleagues. ". . . und nun ging
der Teufel los" ("and now the Devil was let loose") he wrote. His
discovery very soon made an impact on practical medicine. In
physics it led to a chain of new and sensational findings. When
Henri Becquerel was exposing salts of uranium to sunlight to
study whether the new radiation could have a connection with
luminescence, he found out by chance—thanks to a few days of
cloudy weather—that another new type of radiation was being
spontaneously emanated without the salts of uranium having to
be illuminated—a radiation that could pass through metal foil
and darken a photographic plate. The two researchers who were
to play a major role in the continued study of this new radiation
were Marie and Pierre Curie. . . .

Research and Surprising Results

Becquerel's discovery had not aroused very much attention.
When, just a day or so after his discovery, he informed the Monday meeting of *l'Académie des Sciences*, his colleagues listened
politely, then went on to the next item on the agenda. It was
Röntgen's discovery and the possibilities it provided that were
the focus of the interest and enthusiasm of researchers. Becquerel
himself made certain important observations, for instance that
gases through which the rays passed become able to conduct
electricity, but he was soon to leave this field. Marie decided to
make a systematic investigation of the mysterious "uranium
rays". She had an excellent aid at her disposal—an electrometer
for the measurement of weak electrical currents, which was con-
structed by Pierre and his brother.

Results were not long in coming. Just after a few days, Marie
discovered that thorium gives off the same rays as uranium. Her
continued systematic studies of the various chemical compounds
gave the surprising result that the strength of the radiation did not
depend on the compound that was being studied. It depended
only on *the amount* of uranium or thorium. Chemical compounds
of the same element generally have very different chemical and
physical properties: one uranium compound is a dark powder,
another is a transparent yellow crystal, but what was decisive for
the radiation they gave off was only the amount of uranium they
contained. Marie drew the conclusion that the ability to radiate
did not depend on the arrangement of the atoms in a molecule,

it must be linked to the interior of the atom itself. This discovery was absolutely revolutionary. *From a conceptual point of view it is her most important contribution to the development of physics.* She now went through the whole periodic system. Her findings were that only uranium and thorium gave off this radiation.

Marie's next idea, seemingly simple but brilliant, was to study the natural ores that contain uranium and thorium. She obtained samples from geological museums and found that of these ores, pitchblende was four to five times more active than was motivated by the amount of uranium. It was her hypothesis that a new element that was considerably more active than uranium was present in small amounts in the ore.

Collaboration and Hard Work

Fascinating new vistas were opening up. Pierre gave up his research into crystals and symmetry in nature which he was deeply involved in and joined Marie in her project. They found that the strong activity came with the fractions containing bismuth or barium. When Marie continued her analysis of the bismuth fractions, she found that every time she managed to take away an amount of bismuth, a residue with greater activity was left. At the end of June 1898, they had a substance that was about 300 times more strongly active than uranium. In the work they published in July 1898, they write, "We thus believe that the substance that we have extracted from pitchblende contains a metal never known before, akin to bismuth in its analytic properties. If the existence of this new metal is confirmed, we suggest that it should be called *polonium* after the name of the country of origin of onc of us." It was also in this work that they used the term *radioactivity* for the first time. After another few months of work, the Curies informed the *l'Académie des Sciences*, on December 26, 1898, that they had demonstrated strong grounds for having come upon an additional very active substance that behaved chemically almost like pure barium. They suggested the name of *radium* for the new element.

In order to be certain of showing that it was a matter of new elements, the Curies would have to produce them in demonstrable amounts, determine their atomic weight and preferably isolate them. To do so, the Curies would need tons of the costly pitchblende. However, it was known that at the Joachimsthal mine in Bohemia large slag-heaps had been left in the surrounding forests. Marie considered that radium ought to be left in the residue. A sample was sent to them from Bohemia and the slag was found to be even more active than the original mineral. Several tons of pitchblende was later put at their disposal through the good offices of the Austrian Academy of Sciences.

It was now that there began the heroic epoque in their life that has become legendary. At this stage they needed more room, and the principal of the school where Pierre worked once again came to their aid. They could use a large shed which was not occupied. There the very laborious work of separation and analysis began. Marie carried out the chemical separations, Pierre undertook the measurements after each successive step. Physically it was heavy work for Marie. She processed 20 kilos of raw material at a time. First of all she had to clear away pine needles and any perceptible debris, then she had to undertake the work of separation. "Sometimes I had to spend a whole day stirring a boiling mass with a heavy iron rod nearly as big as myself. I would be broken with fatigue at day's end," she writes.

In a preface to Pierre Curie's collected works, Marie describes the shed as having a bituminous floor, and a glass roof which provided incomplete protection against the rain, and where it was like a hothouse in the summer, draughty and cold in the winter; yet it was in that shed that they spent the best and happiest years of their lives. There they could devote themselves to work the livelong day. Sometimes they could not do their processing outdoors, so the noxious gases had to be let out through the open windows. The only furniture were old, worn pine tables where Marie worked with her costly radium fractions. Since they did not have any shelter in which to store their precious products the latter were arranged on tables and boards. Marie could remember the joy they felt when they came into the shed at night, seeing "from all sides the feebly luminous silhouettes" of the products of their work. The dangerous gases of which Marie speaks contained, among other things, radon—the radioactive gas which is a matter of concern to us today since small amounts are emitted from certain kinds of building materials. Wilhelm Ostwald, the highly respected German chemist, who was one of the first to realize the importance of the Curies' research, traveled from Berlin to Paris to see how they worked. Neither Pierre nor Marie was at home. He wrote: "At my earnest request, I was shown the laboratory where radium had been discovered shortly before. . . .

Pierre and Marie Curie (pictured) believed that the results of their hard work should be available to all researchers.

It was a cross between a stable and a potato shed, and if I had not seen the worktable and items of chemical apparatus, I would have thought that I was being played a practical Joke.". . .

The Nobel Prize

In view of the potential for the use of radium in medicine, factories began to be built in the USA for its large-scale production. The question came up of whether or not Marie and Pierre should apply for a patent for the production process. They were both against doing so. Pure research should be carried out for its own sake and must not become mixed up with industry's profit motive. Researchers should be disinterested and make their findings available to everyone. Marie and Pierre were generous in supplying their fellow researchers with the preparations they had so laboriously produced. They furnished industry with descriptions of the production process.

In 1903, Marie and Pierre Curie were awarded half the Nobel Prize in Physics. The citation was, "in recognition of the extraordinary services they have rendered by their joint researches on the radiation phenomena discovered by Professor Henri Becquerel." Henri Becquerel was awarded the other half for his discovery of spontaneous radioactivity.

Marie and Pierre Curie Discover Radium:
December 26, 1898

For the Beauty of Science

by Marie Curie

Marie and Pierre Curie were scientific partners throughout their lives. Just three years after receiving the 1903 Nobel Prize, Pierre was struck and killed by a horse carriage. Though deeply saddened by the loss of her husband, Marie continued to pursue scientific research. In 1908, she became the first woman professor at the Sorbonne in France. She received a second Nobel Prize in 1911 in the field of chemistry for having discovered the chemical properties of radium.

Marie Curie devoted her life to research and had a successful career both with Pierre and after his death. As a woman scientist making important inroads in a field dominated by men, Marie won the admiration of many and was well received by the public. She delivered the following address at Vassar College in 1921. In it she reflects on the progress in radium research in the twenty-three years since its discovery. Beyond that, she relates the conditions surrounding the radium discovery and the ideas that made the discovery possible. Curie discusses the chain of thinking which led her to posit the notion that radioactivity might be an atomic property. After years of hard work, she says, the husband-and-wife team was able to separate the radium element.

Curie also explains what it is that makes radium so important: the effects of its rays. She cites, as an example, the use of radiation in cancer treatment. Although Curie is pleased at the consequences of her discovery, she indicates that the discovery of radium was made for the beauty of science and that the beneficial aspects were merely

---

Marie Curie, address at Vassar College, Poughkeepsie, NY, May 14, 1921.

a welcome by-product. Curie concludes by expressing her hope that others, and perhaps some of those people sitting in the audience, will contribute to the scientific progress of radium research.

I could tell you many things about radium and radioactivity and it would take a long time. But as we cannot do that, I shall only give you a short account of my early work about radium. Radium is no more a baby, it is more than twenty years old, but the conditions of the discovery were somewhat peculiar, and so it is always of interest to remember them and to explain them.

We must go back to the year 1897. Professor Curie and I worked at that time in the laboratory of the school of Physics and Chemistry where Professor Curie held his lectures. I was engaged in some work on uranium rays which had been discovered two years before by Professor [Antoine Henri] Becquerel.

Greater Radioactivity

I spent some time in studying the way of making good measurements of the uranium rays, and then I wanted to know if there were other elements, giving out rays of the same kind. So I took up a work about all known elements, and their compounds, and found that uranium compounds are active and also all thorium compounds, but other elements were not found active, nor were their compounds. As for the uranium and thorium compounds, I found that they were active in proportion to their uranium or thorium content. The more uranium or thorium, the greater the activity, the activity being an atomic property of the elements, uranium and thorium.

Then I took up measurements of minerals and I found that several of those which contain uranium or thorium or both were active. But then the activity was not what I could expect, it was greater than for uranium or thorium compounds like the oxides which are almost entirely composed of these elements.

Then I thought that there should be in the minerals some unknown element having a much greater radioactivity than uranium or thorium. And I wanted to find and to separate that element, and I settled to that work with Professor Curie. We thought it would be done in several weeks or months, but it was not so. It took many years of hard work to finish that task. There was not one new element, there were several of them. But the most important is radium, which could be separated in a pure state.

Intensity of Radium Rays and Medical Treatment

Now, the special interest of radium is in the intensity of its rays which is several million times greater than the uranium rays. And the effects of the rays make the radium so important. If we take a practical point of view, then the most important property of the rays is the production of physiological effects on the cells of the human organism. These effects may be used for the cure of several diseases. Good results have been obtained in many cases. What is considered particularly important is the treatment of cancer. The medical utilization of radium makes it necessary to get that element in sufficient quantities. And so a factory of radium was started to begin with in France, and later in America, where a big quantity of ore named carnotite is available. America does produce many grams of radium every year, but the price is still very high because the quantity of radium contained in the ore is so small. The radium is more than a hundred thousand times dearer than gold.

But we must not forget that when radium was discovered no one knew that it would prove useful in hospitals. The work was one of pure science. And this is a proof that scientific work must not be considered from the point of view of the direct usefulness of it. It must be done for itself, for the beauty of science, and then there is always the chance that a scientific discovery may become like the radium a benefit for humanity.

Hoping for Continued Progress

The scientific history of radium is beautiful. The properties of the rays have been studied very closely. We know that particles are expelled from radium with a very great velocity near to that of the light. We know that the atoms of radium are destroyed by expulsion of these particles, some of which are atoms of helium. And in that way it has been proved that the radioactive elements are constantly disintegrating and that they produce at the end ordinary elements, principally helium and lead. That is, as you see, a theory of transformation of atoms which are not stable, as was believed before, but may undergo spontaneous changes.

Radium is not alone in having these properties. Many having

other radio-elements are known already, the polonium, the mesothorium, the radiothorium, the actinium. We know also radioactive gases, named emanations. There is a great variety of substances and effects in radioactivity. There is always a vast field left to experimentation and I hope that we may have some beautiful progress in the following years. It is my earnest desire that some of you should carry on this scientific work and keep for your ambition the determination to make a permanent contribution to science.

12 The Boxer Rebellion in China Stirs the World: June 20–August 14, 1900

Antiforeign Sentiment in China

by Diana Preston

In the second half of the nineteenth century, China was subjected to Western influence and economic interests. Whereas some Chinese considered relations with the West acceptable and worthwhile, other sectors of the Chinese population opposed foreign spheres of influence. In this atmosphere arose an antiforeign secret society, called the Fists of Righteous Harmony. Inspired by mysticism, nationalism, and martial arts, this peasant reactionary group was nicknamed "the Boxers" by foreigners. The movement eventually grew strong and the Boxers launched aggressive attacks against foreigners and Chinese Christians throughout 1898 and 1899.

The Boxer Rebellion was especially violent between June 20 and August 14, 1900. Supported by the ruling dynasty in China and with 140,000 members, the Boxers seized Peking (Beijing) during this time and overtook the foreigners and Chinese Christians in the city. After a summer of violence and oppression, the Boxer uprising was suppressed through the military intervention of Great Britain, France, Germany, the United States, Russia, and Japan. Tens of thousands of people died as a result of the rebellion.

The following selection by Diana Preston depicts the drama in China between June and August 1900. Preston portrays the atmosphere of violence and fear that existed in Peking that summer as well as the pro-Boxer attitude of the imperial court. In addition, she explores the significance of the event both for China and for the West. In Preston's view, the rebellion was pivotal. Not only did it trigger the eventual collapse of the ruling dynasty, but it also

Diana Preston, *The Boxer Rebellion: The Dramatic Story of China's War on Foreigners That Shook the World in the Summer of 1900*. New York: Walker & Company, 2000. Copyright © 2000 by Walker & Company. Reproduced by permission.

amounted to a lasting disruption of China's relations with the West.

Diana Preston is a historian, writer, and broadcaster. She has written books on a variety of topics.

"Standing together as the sun rose fully, the little remaining band, all Europeans, met death stubbornly. . . . As one man fell others advanced, and finally, overcome by overwhelming odds, every one of the Europeans remaining was put to the sword in a most atrocious manner." So read a dramatic dispatch in the London *Daily Mail* of 16 July 1900 from its special correspondent in Shanghai. Under the headline "The Pekin Massacre," it confirmed in gruesome detail what the world already suspected—that hundreds of foreigners besieged in Peking's diplomatic quarter since 20 June had been murdered.

The Boxer Rising and International Response

The news flew around the world, gaining in horrific detail. The *New York Times* dwelt on the fate of the Russian minister and his wife, plunged into boiling oil. It informed its readers that the besieged "went mad and killed all their women and children with revolvers."

In the event, these reports proved false. They might so easily have been true. The summer of 1900 witnessed a pivotal episode in China's fractured relationship with the West—the Boxer rising. It was an event that left tens of thousands dead and touched the lives of millions more. It precipitated the end of the ruling Manchu dynasty. It tainted China's relationship with the wider world, and continues to do so even today.

The Boxer rising prompted an unprecedented international response. It saw the first steps in America's gradual assumption from Britain of the role of an often misunderstood, sometimes misguided, and occasionally hypocritical world policeman. It brought home to Americans the moral compromises—and the cost in human lives—becoming involved with diversely motivated allies in interventions against hostile populations in distant alien lands. It also revealed Japan's growing confidence and military prowess to a startled world. Soon she would become China's greatest tormentor.

The Boxers themselves were an unlikely catalyst for such far-reaching effects. An obscure, ill-organized sect that claimed to possess supernatural powers, it drew its members mainly from the poor and dispossessed of northern China. The foreigners called them "Boxers" because of the ritualistic martial arts they practiced. Their lives had long been a losing struggle against cycles of flood, drought, and famine. The arrival in China of increasing numbers of foreigners had only deepened their misery. Some foreigners came in pursuit of commerce, and the new technologies they brought with them—steamboats and locomotives, telegraph systems and mining equipment—not only offended the spirits of earth, water, and air but also robbed many Chinese of their jobs. Christian missionaries—fresh-faced and idealistic men and women from the American Midwest, bearded priests from Germany and France—came in search of souls. Often ignorant, dismissive, or contemptuous of the native culture, they and their aggressive proselytizing threatened the very fabric of Chinese family and village life. The Boxers despised their Chinese converts as traitors, "rice Christians" who had sold themselves for a square meal.

An Eruption of Resentment

The Boxers' simmering resentment erupted across the northern provinces of Shantung, Shansi, and Chihli in the summer of 1900. Chanting mobs surrounded the mission stations and dragged out their terrorized occupants. Some they killed on the spot; others they took to Boxer temples to be slowly tortured to death. Tens of thousands of Chinese converts, Protestant and Catholic, were murdered—hacked to pieces, skinned alive, set alight, or buried still living.

The Boxers attacked and murdered Western railway engineers and burned down stations. Reinforced by Imperial Chinese troops, they blockaded 600 foreigners and some 4,000 Chinese Christians in the international port of Tientsin. The foreigners in Peking—nearly 900 men, women, and children from the eighteen most powerful nations in the world—were besieged in the diplomatic quarter. Established in the 1860s following China's defeat in the war with Britain and France, the quarter was by 1900 a commercial as well as a diplomatic district. Banks, shops, and offices prospered alongside the embassies—or "legations," as they were then known—of America, Britain, Russia, Japan, and many of the countries of Western Europe.

The siege forced the quarter's motley population—diplomats and missionaries, academics and adventurers, soldiers and visiting socialites, journalists and engineers—to cooperate despite differences of language and custom and a long history of petty rivalries. The women sewed sandbags of expensive silks, and everyone, from ambassadors' wives to Orthodox priests, filled them. Brave men, both marines and civilian volunteers, fought and died behind them. The cowardly hid in cellars on the flimsiest of excuses. The besieged cooked and ate unappetizing, indigestible meals of rice and horsemeat and were glad of them. Nearly everyone got dysentery. In the heat of the humid Peking summer, thick swarms of black flies and the sickly sweet stench of rotting human flesh were everywhere. The plight of the three thousand or so Chinese converts sheltering in a carefully segregated part of the compound was worst of all. Denied an equal distribution of the food, they were soon reduced to stripping the bark off trees and devouring crows and dogs bloated on human corpses in an effort to survive.

At first everyone listened for the sound of the relief force's guns. When it did not come some men made preparations to shoot their wives and children should the "yellow fiends" overrun the complex. Others smoked cigars and swilled champagne from the well-stocked cellars. A few cracked under the strain and were locked up. The more phlegmatic wondered how they had gotten into a predicament that even the supposed experts had failed to foresee.

In the nearby Peitang Cathedral, a handful of French and Italian guards under the glinting eye of the martial French bishop Favier struggled against frightening odds and in even worse conditions to defend a community of nearly thirty-five hundred souls.

Watching and waiting on events from within Peking's fabled pink-walled Forbidden City was "the Old Buddha," Tzu Hsi, the sixty-five-year-old Empress Dowager of China. In an otherwise totally male-dominated society, this extraordinary woman had held power, directly or indirectly, for nearly forty years. To many of the foreign community she was an Asiatic Catherine de Medici, a woman of unimaginable sexual appetites and political ambition who murdered anyone, including her closest family, who stood in her way. A reactionary to the core, she had recently incarcerated her nephew the emperor for daring to lead a reform movement. She shared the Boxers' loathing of the foreigners in

China and was astute enough to realize two things: First, the Boxers could help her sweep the hated interlopers out of China, and second, their genuine social and economic grievances had to be harnessed or they might be turned against her and the Manchu dynasty. She therefore turned a blind eye to the Boxers' murderous activities, then gave them official support. It proved her greatest mistake.

Foreign Cooperation Ends the Rebellion

Foreign troops relieved Tientsin in July while an international relief force eventually raised the sieges of the diplomatic quarter and the Peitang Cathedral in August. By then, however, over 200 foreigners had been killed or wounded in the diplomatic quarter and hundreds of Chinese Christians had perished, most from starvation and disease. In the Peitang over 400 Chinese and foreigners had died, including 166 children. In the Chinese hinterland, some 200 foreign nuns, priests, and missionaries and their families had been murdered while the death toll of Chinese converts ran into tens of thousands. The numbers of Boxers and Imperial troops who perished can only be guessed.

Both Tientsin and Peking were thoroughly and indiscriminately looted by all nationalities and all classes. Many innocent civilians committed suicide rather than face being raped and killed. Moats, rivers, and wells became clogged with bodies. It had been a terrible bloodbath and it had backfired badly on Tzu Hsi. In her final years she was forced to embrace the very reforms she had earlier resisted and indeed to introduce them so quickly that they undermined the dynasty she had fought to preserve. The last emperor, Pu Yi, was deposed in 1911, just three years after her death.

It was an irony that the foreign powers that had cooperated to defeat China would soon be at war with one another. Indeed, the barely suppressed rivalries among them during the crisis in China had signaled to many the inevitability of an impending major conflict. Yet the allied action against the Boxer rising also set an important precedent, one that prefigures the relief and policing activities of the United Nations and NATO [North Atlantic Treaty Organization]. It certainly caught the imagination of the many contemporaries who likened it to the Crusades. It had been, as British Prime Minister Lord Salisbury put it, "an entirely new experiment." Nevertheless, it would take two world wars and one cold one to achieve such a widespread collaboration again.

The Boxer Rebellion was an extraordinary event—heroic and farcical, tragic and shocking, brutal and ridiculous, with far-reaching implications. Yet it was also a richly human story. The foreigners who survived wrote a great deal about the rising and its aftermath. Articulate, opinionated, often prejudiced, their accounts provide fascinating insight into what occurred during those traumatic months.

The Boxer Rebellion in China Stirs the World: June 20–August 14, 1900

A Missionary in the Midst of the Uprising

by Eva Jane Price

Many of the foreigners who were targets of the Boxer Rebellion between June 20 and August 14, 1900, were missionaries. In the late nineteenth century many Christian missionaries from the West journeyed to China to spread Christianity. Since this religion was not native to China, missionaries were a symbol of unwanted foreignness to the Boxers.

One missionary family that became caught up in the events of the summer of 1900 was that of Charles and Eva Jane Price. The Prices, along with their two children, had been in China since their 1889 arrival from Oberlin College in Ohio. The following letter of June 29, 1900, provides an eyewitness account of the Boxer Rebellion by Eva Jane Price, whose family was trapped in the siege of June 20 through August 14. In this account, Price reveals the anxiety of Boxer oppression. She relates the experiences of neighbors subjected to looting and threats. She also recounts the investigation of a murder and its impact on the community.

The Prices became victims of the Boxer Rebellion. Eva Jane and her family, along with several other missionary families, were slain by the Boxers on August 15, 1900.

J ust what we have learned that I was going to write I shall never remember, I fear, for since the words were written we have learned what it is to meet danger, to have the feeling that we were facing death itself, and to realize as never before the power and majesty of our great Heavenly Father. Two weeks ago friends came to us for a visit partly, and partly to be in a safer place. Miss Eldred came from the district where the greatest outbreak has been, Hung T'ung, and Mr. and Mrs. Lundgren from Chieh Hsiu, sixty li [1 li = ⅓ mile] away. Since they have been here we have thought and been able to talk of little besides the dangers around us. We tried to even pass a rule that we would not talk about it, but our hearts have been too full. Often just at nightfall disturbing letters or rumors have come in until for several nights we have not slept as usual. Last Sunday the most distressing rumors came that all the foreigners in Peking had been killed. Our mail has been shut off, so we have not been able to learn anything authentic. Later in the day the Kuan [local magistrate] posted a notice in favor of the "Boxers." Again, later, he put up another saying foreigners must be respected. Every day since there have been such conflicting rumors that one has been all but dissected. Our hearts have grown faint one hour to be lifted again the next.

The Neighbors' Experience

On Wednesday evening we took the precaution to pack two trunks with most necessary things, and the next day they were secretly buried in the chicken house with a box containing most of our money. We seemed to be momentarily expecting an attack in which we would at least be robbed if nothing worse. To put out of sight the most necessary things seemed wise. This was done without the knowledge of the servants so far as we know.

Last eve we sat out in the court as usual trying to be brave and forget the situation. I went in about eight to help Florence get ready for bed. Just as we had finished, Miss Eldred passed hastily along the hall and called out in a quiet voice, "Mrs. Price." I answered as quietly and followed her into her room. "They have come. They are over at Mr. Atwater's now." I went back and had Florence dress again, but she only thought it a funny freak. We packed another box of clothes that would be needed if we were not harmed and went about as quietly as possible, fearing to hear from Mr. Atwater's. A man came shortly with the word that the

Atwaters had fled to the *yamen* [a government office; here it refers to the magistrate's office], and their house was being looted. "They will be here before very long then," was our only thought, and no one can imagine the terrible suspense of the next hour as we went about dreading we knew not what. For days we had lived not knowing what an hour would bring forth. We heard Dr. Atwood's house and the chapel were to be destroyed. The "nightshirt brigade" formed and sallied out one night on a false alarm, but now that the tension was broken we expected the worst. The servants bound up their heads with dark cloth to prevent being seen and to get their queues where they would not be easily taken hold of. They sharpened case knives as weapons of defense. The "nightshirt brigade" put on revolvers, gun, rifle, but in none of these things could we trust. We kept crying to God to deliver us. Definite word came soon that the Kuan had furnished an escort for the Atwaters, and they were back in their home. A note came from them telling briefly of their experience, and I began an answer expressing our sympathy. But before I had time, I heard voices and knew Mrs. Atwater and the little girls had come. It was 10:30 and, getting the little ones down as quickly as possible, we then listened to a correct version of the affair.

She said a pounding and yelling began while they were at supper and went on until a large crowd had gathered. Bricks and pieces of tile came flying over the walls into their court, one of which came near killing Celia, another just missed Mr. Atwater's head. During the suspense they could hear blow after blow on their small side gate, which seemed the chief point of attack. There were sounds of chopping going on persistently with the other pounding and yelling, and just as the gate gave way, their cook came and urged them to go out the front way to the *yamen*. To open the front gate in face of hundreds of those come to harm you and to walk boldly among them must have required divine courage. On they went untouched, stumbling over a strange road, through dust, over rough places in the dark, feeling that escaping for life is an experience one need never crave, but Mrs. Atwater was enabled to endure it all only to find themselves shut out of the *yamen* itself when they reached it. They were made to stay in the outer court into which gathered a motley crowd. Mrs. Atwater said that it was good as a play to sit there on a narrow hard bench holding one of the children and watching what went on around. The weird light of flashing lanterns, the loud voices of the excited crowd, the barring and guarding of the door that led into the Kuan's court. There were signs of hurrying to and fro, and the chair of state used by the Kuan was carried out guarded by soldiers but empty as we learned afterward, for the Kuan at the very first notice of the affair had rushed out to the scene of action not even waiting to put on his official robes or for his chair, which he ordered to come after him. We have the greatest admiration for an official who himself went into the thickest of the mob and with his own hands took a man making off with a rug under his arm.

While the Atwaters sat there in the courtyard of the *yamen*, four prisoners were brought in howling in most abject fear. The brave invulnerable "Boxer" met his match in Fen Cho fu's brave official. He dispersed the crowd in a few moments after his arrival, but they had had time to smash windows, stoves, dishes, lamps, etc., which were strewn all over the court. When the Kuan went back to the *yamen*, the Atwaters were sent back with an escort. They walked through the courts over broken pieces of what had been their housekeeping outfit. Chairs, tables, windows, dishes, etc., however, owing to the shortness of time the mob had for its work, were the chief things destroyed. Some other things were stolen. What a home to go back to at ten o'clock at night: two little children tired, excited, and still fearing another mob! They decided to come here. They felt it was running a gauntlet to do so, but it seemed best, and they set off once more in the dark and dust, feeling it a question whether they could reach us alive. The "boys" helped valiantly during the whole disturbance, and we have great reason to be proud of them and their bravery. The necessities were carried over here, and we quieted down about midnight. Mr. Atwater went back and slept in their demoralized home; fortunately there had not been time to get into the bedrooms, and they were in their usual order. Sleep can never be refreshing under such circumstances, and there was still the feeling of doubt as to whether the Kuan would punish the prisoners in anything like an adequate manner. Unless he did, we were only at the mercy of the mob that could gather at any time.

A Murder Is Investigated

After breakfast yesterday morning our anxieties began afresh, and about half an hour afterward, to add to the horror of our feelings, word came in that a foreigner had been murdered about ten li from the city. We began to go through the dreadful possibilities as to who it could be. Mr. Davis was the first one to come into our minds; then maybe it was Mr. Williams or Mr. Jennings or Mr. Ogren. There were good reasons for serious doubt as to who the murdered man could be, whether it could be a foreigner at all, but we were in a condition of mind to believe the worst. Some foreigner, we could not know who, had evidently been trying to reach us and had lost his life. One could picture details distressing in the extreme and which depressed us greatly in our already strained frame of mind. For hours we were in suspense. Word came that the Kuan had sent out to have the matter investigated. Awful complications presented themselves. Later, the encouraging word came that the Kuan had himself asked Mr. Han, our evangelist, to go with him to help identify the nationality of the man; and we waited in great anxiety. It seems now as though God himself came to the earth to thwart the designs of evil men. This is his story as near as I can tell the details. The Kuan had him put on an official hat, put him in a cart with soldiers on either side, and they set out for the village where the murder occurred. On arrival the Kuan called together the headmen of the village and those of several other villages who had congregated and put them through a set of questions: "Was the man a foreigner? Why was he killed?" "He was found poisoning wells, and we killed him." "Yes, well where is the body?"

They were taken to the place where it had been buried. It was dug up and carried into a temple. The Kuan motioned for Mr. Han to approach it first. He did so in great fear, for at first glance his heart sank as he thought it a foreigner. The body was so hacked and cut up as to scarcely hold together. The face was mutilated, but there was something foreign in his appearance and about his dress which did not seem like the usual dress of the natives. The sleeves were a tighter cut and differed somewhat in detail. Mr. Han carefully examined the external and then opened the outer garment to look at the body inside. At first glance he said, "This is not a foreigner." "What!" cried the Kuan, no doubt in great relief. "No, this is no foreigner. Foreigners are clean. Look at this man's clothes; how dirty they are!" "Yes," said the Kuan excitedly. "Let us look for other things." "Look at the fingernails," said Mr. Han. "The foreigners trim their nails. This man's nails are long. And his hair! This isn't a foreigner's hair. This is a Chinaman's hair!" "Yes," said the Kuan. "And the man is an opium eater. Look at his teeth; how black they are. And the edges of his sleeves are stained with opium too—this man can't be a foreigner." Then, like a flash, he turned on the village elders. "What do you mean by saying this man is a foreigner. This is one of our own men!"

The Kuan Takes Action

Then followed a scene—raging and storming and raving. During the fray some women came storming up claiming to belong to the invulnerable "Boxers" making their passes and thrusts in the face of the very Kuan himself. There were three of them probably belonging to one family. The Kuan immediately ordered them to kneel, but they refused. In a rage he kicked one of them over and ordered the soldiers to bring the women and those implicated in the murder to the *yamen*. It was a busy day for the poor Kuan. What with the affair of the night before which he had so heroically put an end to; the prisoners, punishments, having proclamations put up on our places, and on top of it all this awful mysterious murder with the insulting acts of the villagers. It must be made a precedent, so he stormed and waved and raged, broke his own teapots and cups—all probably to make an impression on those around as he had done while out in the village when he threw the teapot, contents and all, directly into the face of the village elder.

Late in the morning we heard what punishment had been meted out to the three women. They were dragged all over the city by their hair around the different suburbs, beaten with thorns, 300 blows on their backs, 40 across their mouths, and they were finally glad to kneel and recant. The man found to be the murderer of the unknown man was beaten and pounded with a spade over the head and face until a sight to behold and then was led away to be kept for execution. Four others implicated had each 1,000 blows. The four men arrested at Mr. Atwater's were beaten and fined different sums of money. The one the Kuan himself captured had a wooden collar fastened around his neck, a board two inches thick with a hole in the middle for the neck. The size of the board was about two feet square, and hinged so it closed around the neck and then locked. The culprit wearing this is said to be made to sit in front of Atwater's place every day so long as the Kuan directs.

CHRONOLOGY

1881
July 2: U.S. president James A. Garfield is shot by Charles Guiteau.
September 19: President Garfield dies from complications of his bullet wounds.

1882
May 20: Germany, Austria-Hungary, and Italy sign the Triple Alliance pact. The pact is renewed every five years until the outbreak of World War I in 1914.
June 30: Charles Guiteau is hanged for having assassinated President Garfield.

1883
January 16: Congress passes the Pendleton Act, which reforms the civil service by regulating federal civil service positions and requiring candidates to pass exams.

1884
October 13–22: The International Meridian Conference, with delegates from twenty-five countries, convenes to establish a universal standard of time.
November 15: The Berlin Conference convenes to set guidelines among the European powers for the partitioning of Africa.

1885
February 26: Participants in the Berlin Conference sign the agreements concerning the division of Africa.
February 27: The Berlin Conference concludes.
July 6: Louis Pasteur successfully uses the rabies vaccine on a young boy named Joseph Meister.

1886
October 28: The Statue of Liberty is dedicated.

1887

Great Britain's Queen Victoria celebrates her golden jubilee (her fiftieth year as reigning queen).

1888

Heinrich Hertz discovers radio waves.
November 15: The Pasteur Institute opens in Paris for the purpose of research into infectious diseases and immunology.

1889

The Eiffel Tower is built in Paris.
February 11: Japan presents a new constitution that explains the role of the emperor, the rights and duties of subjects, and the function of the legislature.
April 20: Adolf Hitler, future leader of the Nazi Party and the Third Reich (1933–1945), is born.

1893

Henry Ford builds his first car.

1894

August 1: The Sino-Japanese War begins over a dispute concerning Korea.
October 15: Captain Alfred Dreyfus is arrested for treason.
December 19–22: The Dreyfus trial takes place. The guilty verdict sparks years of controversy over justice and anti-Semitism in France.

1895

April 17: The Treaty of Shimonoseki ends the Sino-Japanese war, which Japan won.
September 28: Louis Pasteur dies.
November 8: Wilhelm Röntgen discovers X-rays.
December 28: At the Grand Café in Paris, Auguste and Louis Lumière present the first public showing of a motion picture.

1896

Antoine Henri Becquerel discovers radioactivity—the emission of radiation—in uranium.

Guglielmo Marconi receives a patent for his invention, the wireless telegraph.

April 22–23: The United States and Spain declare war upon one another, two months after the *Maine* explodes in Havana Harbor, Cuba.

December 1897–January 1898: Émile Zola publishes letters reprimanding the French government for the injustice of the Dreyfus Affair.

1898

July 18: Marie and Pierre Curie's discovery of the radioactive element polonium is presented to the scientific community by the French Academy of Sciences.

December 10: The Treaty of Paris ends the Spanish-American War. The United States won.

December 26: The French Academy of Sciences announces the discovery of radium, a highly radioactive element, by the Curies.

1900

June 20–August 14: The Boxer Rebellion, both antiforeign and violent in nature, takes place in China.

August 15: Missionaries Charles and Eva Jane Price and their family are slain by the Boxers.

FOR FURTHER RESEARCH

Books

George Akita, *Foundations of Constitutional Government in Modern Japan, 1868–1900*. Cambridge, MA: Harvard University Press, 1967.

William Ralston Balch, *The Life of James Abram Garfield, Late President of the United States*. Philadelphia: Hubbard Bros., 1881.

George Frederick Barker, *Röntgen Rays: Memoirs by Röntgen, Stokes, and J.J. Thomson*. New York: Harper & Bros., 1899.

Ian Bartky, *Selling the True Time: Nineteenth-Century Time Keeping in America*. Palo Alto, CA: Stanford University Press, 2000.

Otto von Bismarck, *Bismarck: The Man and the Statesman*. New York: Harper & Bros., 1898.

Alan Ralph Bleich, *The Story of X-Rays, from Röntgen to Isotopes*. New York: Dover, 1960.

Jean-Denis Bredin, *The Affair: The Case of Alfred Dreyfus*. Trans. Jeffrey Mehlman. New York: W.W. Norton, 1986.

Michael Burns, *France and the Dreyfus Affair*. New York: Bedford/St. Martin's, 2000.

Eric Cahm, *The Dreyfus Affair in French Society and Politics*. New York: Longman, 1996.

David Carr, *Time, Narrative, and History*. Bloomington: Indiana University Press, 1986.

Sean Dennis Cashman, *America in the Gilded Age: From the Death of Lincoln to the Rise of Theodore Roosevelt*. New York: New York University Press, 1984.

James Chapman, *Cinemas of the World: Film and Society from 1895 to the Present.* London: Reaktion Books, 2003.

Wen Ching, *The Chinese Crisis from Within.* London: Grant Richards, 1901.

Paul A. Cohen, *History in Three Keys.* New York: Columbia University Press, 1997.

Gordon A. Craig, *Germany, 1866–1945.* New York: Oxford University Press, 1978.

Francesco Crispi, *The Memoirs of Francesco Crispi, Vol. III, the Triple Alliance.* Trans. Mary Prichard-Agnetti from documents collected and edited by Thomas Palamenghi-Crispi. New York: Hodder and Stoughton, 1912.

L. Perry Curtis, *Coercion and Conciliation in Ireland, 1880–1892: A Study in Conservative Unionism.* Princeton, NJ: Princeton University Press, 1963.

Justus D. Doenecke, *The Presidencies of James A. Garfield and Chester A. Arthur.* Lawrence: Regents Press of Kansas, 1981.

John W. Dower, *Japan in War and Peace: Selected Essays.* New York: New Press, 1993.

René Dubos, *Pasteur and Modern Science.* Madison, WI: Science Tech, 1988.

Émile Duclaux, *Pasteur: The History of a Mind.* Metuchen, NJ: Scarecrow, 1973.

Jack C. Ellis, *A History of Film.* Boston: Allyn and Bacon, 1995.

Frantz Fanon, *The Wretched of the Earth.* Berkeley, CA: Grove, 1986.

Stig Förster, Wolfgang J. Mommsen, and Ronald Robinson, *Bismarck, Europe, and Africa: The Berlin Africa Conference 1884–1885 and the Onset of Partition.* London: Oxford University Press, 1988.

Takashi Fujitani, *Splendid Monarchy: Power and Pageantry in Modern Japan.* Berkeley: University of California Press, 1996.

James A. Garfield and Charles E. Henry, *Politics and Patronage in the Gilded Age: The Correspondence of James A. Garfield and Charles E. Henry.* Eds. James D. Norris and Arthur H. Shaffer. Madison: State Historical Society of Wisconsin, 1970.

Imanuel Geiss, *German Foreign Policy, 1871–1914.* Boston: Routledge, 1976.

Robert Gellately, *Backing Hitler: Consent and Coercion in Nazi Germany.* Oxford: Oxford University Press, 2001.

Carol Gluck, *Japan's Modern Myths: Ideology in the Late Meiji Period.* Princeton, NJ: Princeton University Press, 1985.

Joseph Goebbels, *The Goebbels Diaries.* Trans. and ed. Louis P. Lochner. London: H. Hamilton, 1948.

George Parkin Grant, *Time as History.* Buffalo, NY: University of Toronto Press, 1995.

J.L. Hammond, *Gladstone and the Irish Nation.* New York: Longman, 1938.

Derek Howse, *Greenwich Time and the Discovery of the Longitude.* New York: Oxford University Press, 1980.

Roger S. Jones, *Physics for the Rest of Us.* Columbus, OH: McGraw-Hill/Contemporary Books, 1993.

Robert Kee, *The Green Flag: The Turbulent History of the Irish National Movement.* New York: Delacorte, 1972.

Stephen Kern, *The Culture of Time and Space, 1800–1918.* Cambridge, MA: Harvard University Press, 1983.

Ian Kershaw, *Hitler, 1889–1936: Hubris.* New York: W.W. Norton, 1999.

John Lukacs, *The Hitler of History.* New York: Vintage Books, 1997.

Auguste Lumière and Louis Lumière, *Letters: Auguste and Louis Lumière.* Eds. Jacques Rittaud-Hutinet and Yvelise Dentzer. Trans. Pierre Hodgson London. Boston: Faber and Faber, 1995.

Lois N. Magner, *A History of the Life Sciences.* New York: M. Dekker, 1994.

James L. McClain, *Japan: A Modern History.* New York: W.W. Norton, 2002.

Albert Memmi, *The Colonizer and the Colonized.* Boston: Beacon, 1991.

W. Robert Nitske, *The Life of Wilhelm Conrad Röntgen, Discoverer of the X-Ray.* Tucson: University of Arizona Press, 1971.

Thomas Pakenham, *The Scramble for Africa, 1876–1912.* New York: Random House, 1991.

John Howard Parnell, *Charles Stewart Parnell: A Memoir by His Brother.* London: Constable, 1921.

Allan Peskin, *Garfield: A Biography.* Kent, OH: Kent State University Press, 1978.

Otto Pflanze, *Bismarck and the Development of Germany.* Princeton, NJ: Princeton University Press, 1990.

Roy Porter, *The Greatest Benefit to Mankind: A Medical History of Humanity.* New York: W.W. Norton, 1997.

Diana Preston, *The Boxer Rebellion: The Dramatic Story of China's War on Foreigners That Shook the World in the Summer of 1900.* New York: Walker, 2000.

Eva Jane Price, *China Journal, 1889–1900: An American Missionary Family During the Boxer Rebellion.* New York: Charles Scribner's Sons, 1989.

Eric Rhode, *A History of the Cinema: From Its Origins to 1970.* London: Allen Lane, 1976.

Emilio Segrè, *From X-Rays to Quarks: Modern Physicists and Their Discoveries.* San Francisco: W.H. Freeman, 1980.

Robert Sklar, *Film: An International History of the Medium.* New York: H.N. Abrams, 1993.

Jeremy Smith, *Britain and Ireland: From Home Rule to Independence.* New York: Longman, 2000.

Thomas Sowell, *Conquests and Cultures: An International History.* New York: BasicBooks, 1998.

Chester C. Tan, *The Boxer Catastrophe.* New York: Columbia University Press, 1955.

René Vallery-Radot, *The Life of Pasteur.* Trans. R.L. Devonshire. New York: Doubleday, 1927.

Gordon Wright, *France in Modern Times: From the Enlightenment to the Present.* New York: W.W. Norton, 1981.

Émile Zola, *The Dreyfus Affair: "J'accuse" and Other Writings.*
Ed. Alain Pagès. Trans. Eleanor Levieux. New Haven, CT:
Yale University Press, 1996.

Websites

Internet Modern History Sourcebook, www.fordham.edu/halsall/
mod/modsbook.html. This outstanding resource for primary
documents presents modern history via outlines and key
word links. The Internet Modern History Sourcebook covers
a variety of subjects and perspectives in world history.

Web Chronology Project from North Park University, http://
campus.northpark.edu/history/WebChron/Introduction.html.
This project from the history department at North Park University in Chicago consists of extensive chronology outlines
for a multitude of historical subjects across many cultures.
An excellent resource, the chronology project also includes
occasional primary and secondary source links.

World History: HyperHistory, www.hyperhistory.com/online_n2/
History_n2/a.html. HyperHistory offers a thorough, user-
friendly series of time lines for people, places, and events in
world history. This is a fine resource for gauging contemporary and simultaneous events in history.

INDEX